The Making of English Popular Culture

The Making of English Popular Culture provides an account of the making of popular culture in the nineteenth century.

While a form of what we might describe as popular culture existed before this period, John Storey has assembled a collection that demonstrates how what we now think of as popular culture first emerged as a result of the enormous changes that accompanied the industrial revolution. Particularly significant are the technological changes that made the production of new forms of culture possible, and the concentration of people in urban areas that created significant audiences for this new culture.

Consisting of fourteen original chapters that cover diverse topics ranging from seaside holidays and the invention of Christmas tradition, to advertising, music and popular fiction, the collection aims to enhance our understanding of the relationship between culture and power, as explored through areas such as 'race', ethnicity, class, sexuality and gender. It also aims to encourage within cultural studies a renewed historical sense when engaging critically with popular culture by exploring the historical conditions surrounding the existence of popular texts and practices.

Written in a highly accessible style *The Making of English Popular Culture* is an ideal text for undergraduates studying cultural and media studies, literary studies, cultural history and visual culture.

John Storey is Emeritus Professor of Cultural Studies at the Centre for Research in Media and Cultural Studies, University of Sunderland, UK. He has published extensively in the field of cultural studies, including ten books, the most recent being the seventh edition of *Cultural Theory and Popular Culture: An Introduction* (2015). His work has been translated into Arabic, Chinese, Dutch, German, Greek, Indonesian, Japanese, Korean, Persian, Polish, Portuguese (Brazil and Portugal), Russian, Serbian, Slovenian, Spanish, Swedish, Turkish and Ukrainian. He is also on editorial/advisory boards in Australia, Canada, China, Germany, Lithuania, the Netherlands, Spain, the UK and the USA, and has been a Visiting Professor at the Technical University of Dresden, the University of Henan, the University of Vienna and the University of Wuhan.

Directions in Cultural History

Series Editors: Gillian Swanson and Ben Highmore

The *Directions in Cultural History* series directs history towards the study of feelings, experiences and everyday habits. By attending to the world of sensation, imagination, and desire at moments of change, and by coupling this to the materials and technologies of culture, it promotes cultural history as a lively and vivid arena for research. The series will present innovative cultural history in an accessible form to both scholars and upper level students.

Print Culture
Frances Robertson

Dreams and Modernity: A Cultural History
Helen Groth and Natalya Lusty

Design at Home: Domestic Advice Books in Britain and the USA since 1945
Grace Lees-Maffei

The Making of English Popular Culture
John Storey

The Making of English Popular Culture

Edited by John Storey

Routledge
Taylor & Francis Group
LONDON AND NEW YORK

First published 2016
by Routledge
2 Park Square, Milton Park, Abingdon, Oxon OX14 4RN

and by Routledge
711 Third Avenue, New York, NY 10017

Routledge is an imprint of the Taylor & Francis Group, an informa business

British Library Cataloguing-in-Publication Data
A catalogue record for this book is available from the British Library

Library of Congress Cataloging-in-Publication Data
Names: Storey, John, 1950– editor.
Title: The making of English popular culture / edited and introduced by John Storey.
Description: Milton Park, Abingdon, Oxon; New York: Routledge, 2016. | Series: Directions in cultural history | Includes bibliographical references and index.
Identifiers: LCCN 2015041192| ISBN 9781138854901 (hardback) | ISBN 9781138854918 (paperback) | ISBN 9781315720678 (ebook)
Subjects: LCSH: Popular culture–England–History–19th century. | England–Social life and customs–19th century. | England–Social conditions–19th century. | England–Intellectual life–19th century. | Social change–England–History–19th century. | Technological innovations–Social aspects–England–History–19th century. | Urbanization–Social aspects–England–History–19th century. | Power (Social sciences)–England–History–19th century.
Classification: LCC DA110 .M153 2016 | DDC 306.0942/09034–dc23
LC record available at http://lccn.loc.gov/2015041192

ISBN: 978-1-138-85490-1 (hbk)
ISBN: 978-1-138-85491-8 (pbk)
ISBN: 978-1-315-72067-8 (ebk)

Typeset in Sabon
by HWA Text and Data Management, London

For Charlie

Contents

Figures

Contributors

Allison Cavanagh is a Lecturer in Communications Studies at the University of Leeds, UK. She teaches and researches in the history of promotional communication and in nineteenth and early twentieth century media. Her most recent article on mediatisation in the nineteenth century, "Barbarous cruelty at the British Museum", explored class identities and the social and cultural uses of Victorian newspapers.

Ralph Crane is Professor and Head of English at the University of Tasmania, Australia. He has published widely on colonial and postcolonial fictions, and has written or edited twenty books. His recent work includes several publications in the area of island studies and the geohumanities. With Lisa Fletcher he is the co-author of *Cave: Nature and Culture*, as well as several articles and book chapters on imperial adventure fiction, illustration and popular fiction in the Victorian Age, and the geohumanities.

Sarah Edge is a Professor of Photography and Cultural Studies at the University of Ulster, Northern Ireland. Her articles on early photography include: 'The Great Famine: absence, memory and photography' in *Cultural Studies* 24: 6, with Gail Baylis, 'Photographic History and the Visual Appearance of An Irish Nationalist Discourse 1840–1870', in *Victorian Literature and Culture* 23: 1 and 'The Power to Fix the Gaze: Gender and Class in Victorian Photography of Pit-Brow Women', in *Visual Sociology* 12: 2. She is currently writing, 'The Extraordinary Archive of Arthur J. Munby: Photographing Class and Gender in the Nineteenth Century' for publication by I.B. Tauris in 2016. She is also an art photographer and regularly exhibits her work.

Lisa Fletcher is Senior Lecturer in English at the University of Tasmania, Australia. She is the author of *Historical Romance Fiction: Heterosexuality and Performativity* (2008) and numerous journal articles and book chapters about islands and literature. With Ralph Crane she is the co-author of *Cave: Nature and Culture*, as well as several articles and book chapters on imperial adventure fiction, illustration and popular fiction in the Victorian Age, and the geohumanities.

John Paul Green is Principal Lecturer at the University of Sunderland, UK. His research interests include science fiction and horror television and cinema, media history and fictional British heroes and national identity. He has published on *Doctor Who* in 'The Regeneration Game: The Changing Faces of Heroism' (2010) and James Bond, with 'You Know My Name, But Who Am I? Structure and Agency in the Making and Remaking of James Bond' (2008) and 'From Bringland With Love' (2007). He has contributed to the BBC television series *Arena* and *Inside Out* as well as appearing in three episodes of *Doctor Who* (2006).

David Huxley is a Senior Lecturer on the BA (Hons) Film and Media Studies course at Manchester Metropolitan University, UK. His subject specialisms and research interests are the graphic novel, comic book and comic strip, censorship, Hollywood film and animation and British music hall performers. He is the author of *Nasty Tales: British Underground Comics, Critical Vision, Manchester*, and has written on British music hall in the *Journal of Early Popular Visual Culture* and the *Journal of Comedy Studies.* He is the editor of the *Journal of Graphic Novels and Comics.*

David James is a Senior Lecturer on the BA (Hons) Film and Media Studies course at Manchester Metropolitan University, UK. His research interests include the portrayal of class in British wartime films, British war films of the 1950s and in British sitcoms of the 1960s, 1970s and 1980s. He was awarded a PhD by Manchester Metropolitan University in March 2006, for his thesis on the representation of class in the sitcoms of Perry and Croft. In conjunction with David Huxley, his current research is based around the question of how music hall comedians and comedy made the transition to film in the UK.

Jochen Petzold studied in Freiburg and Eugene, Oregon, USA. His PhD project focussed on contemporary South African literature, his second book (Habilitation) examines speech situations in lyrical poetry. From 2009 he was Marie Curie Fellow (ERC funded) at the Institute for Advanced Studies in the Humanities at the University of Edinburgh, Scotland, examining the portrayal of natural science in Victorian juvenile magazines. In 2010, he was appointed Professor of British Studies at the University of Regensburg, Germany. His current research interests include South African literature and Victorian popular culture, particularly juvenile literature.

Elodie A. Roy received her doctorate from the International Centre for Music Studies, Newcastle University, UK in 2014. Her work examines the cultural history of phonography and the shifting relationships between music objects and memory practices from the late nineteenth century onwards. She has written articles and book chapters in the areas of popular music, media theory and twentieth-century literature. Her first monograph, entitled *Media, Materiality and Memory: Grounding the Groove*, was published by Ashgate (Music and Material Culture series) in October 2015. In spring 2016, she was awarded a short postdoctoral fellowship from the Centre Marc Bloch, Berlin.

Ralf Schneider is Professor of British Literary and Cultural Studies at Bielefeld University, Germany. He has worked and published on the theory and analysis of literary character, cognitive narratology and a variety of topics in British literary and cultural history, including the cultural memory of the First World War, war and British identities, popular poetry anthologies and masculinities in nineteenth-century prose fiction. He is currently working on the narrative representation of migration in the contemporary British novel and on a handbook of British literature and culture of the First World War (co-edited with Jane Potter).

Robert Shail is Professor of Film at the Northern Film School, Leeds Beckett University, UK. He has published widely on postwar British cinema and his study of the Children's Film Foundation is due for publication shortly from Palgrave-Macmillan/BFI. His recent research work has also branched out into the history of British popular culture, focusing particularly on comic books and including an essay on Beryl the Peril for the *Journal of Graphic Novels and Comics.*

Jenny Skipp is a Senior Lecturer at the University of Trier, Germany, teaching English language and cultural studies, and wors at the University of Luxembourg teaching English for Academic Purposes. She received her PhD in 2007 from Leeds University, UK for a thesis which quantitatively and qualitatively reassessed British eighteenth-century erotic literature. Since then she has written articles and given papers on themes such as erotic culture, manhood and masculinity, and the representation of female sexuality.

John Storey is Emeritus Professor of Cultural Studies at the Centre for Research in Media and Cultural Studies, University of Sunderland, UK. He has published extensively in the field of cultural studies, including ten books, the most recent being the seventh edition of *Cultural Theory and Popular Culture: An Introduction* (2015). His work has been translated into Arabic, Chinese, Dutch, German, Greek, Indonesian, Japanese, Korean, Persian, Polish, Portuguese (Brazil and Portugal), Russian, Serbian, Slovenian, Spanish, Swedish, Turkish and Ukrainian. He is also on editorial/advisory boards in Australia, Canada, China, Germany, Lithuania, the Netherlands, Spain, the UK and the USA, and has been a Visiting Professor at the Technical University of Dresden, Germany, the

University of Henan, China, the University of Vienna, Austria and the University of Wuhan, China.

Robert Troschitz is Lecturer in British Cultural Studies at the Technical University of Dresden, Germany. He recently finished his PhD with a thesis on 'The Idea of English Higher Education and the Concept of the Student (1943–2011)'. He has published and given talks on student activism, neoliberal education policies and the history of popular culture, in particular holiday making.

Claire Westall is a Lecturer in the Department of English and Related Literature at York University, UK. Her work investigates the literary-cultural legacies of British imperialism, in relation to Caribbean literature and wider questions of postcolonial/global/world literary studies, and in relation to debates about England, Englishness and devolution in Britain. She has a range of publications on cricket and Caribbean literature and is co-editor of *Cross-Gendered Literary Voices: Appropriating, Resisting, Embracing* (2012), and co-editor of *Literature of an Independent England: Revisions of England, Englishness and English Literature* (2013). Claire is also co-author of *The Public on the Public: The British Public as Trust, Reflexivity and Political Foreclosure* (2015).

Anne-Julia Zwierlein is Chair of English Literature and Culture at the University of Regensburg, Germany. She specialises in Victorian and early modern culture. Among her publications are the monographs *Majestick Milton: British Imperial Expansion and Transformations of Paradise Lost, 1667–1837* (2001) and *Physiology and the Nineteenth-Century Novel of Formation* (2009), as well as (co-)edited volumes on interiority (2002), Jacobean city comedy (2004), nineteenth-century biology and culture (2005), *Gender and Creation* (2010), *Interdisciplinary Perspectives on Aging in Nineteenth-Century Culture* (2013), *Gender and Disease in Literary and Medical Cultures* (2014), and a special issue of *Zeitschrift für Anglistik und Amerikanistik (ZAA)* on *Victorian Oral Cultures* (2015). She is currently engaged in a project on Victorian 'oral cultures' and mass literacy (funded by the German Research Foundation, DFG).

Introduction

Making popular culture

John Storey

The first aim of this book is to provide an account of the making of English popular culture in the nineteenth century. The book will not argue that something we might describe as popular culture did not exist before this period, but it will present evidence to show that what we now think of as popular culture, in all its contradictory meanings, first emerges during the enormous changes that accompanied what we call the industrial revolution. In other words, contemporary popular culture, in terms of concepts, texts and practices, has its foundations in the nineteenth century. Of particular importance are the technological changes that made the production of new forms of culture possible and the concentration of people in urban areas that created significant audiences for this new culture. But, as the title of this book indicates, popular culture had to be 'made', and by the 'making' of popular culture the chapters collected here mean two things: 1) what was produced by the new culture industries; and 2) how these new productions were consumed by the new urban audiences.

The second aim of the book is to encourage within cultural studies a renewed historical sense when engaging critically with popular culture. Cultural studies has never been quite as historical in its approach to culture as it has often claimed to be. Despite the fact that one of its 'founding fathers' was the historian E.P. Thompson, too often it has taken overly seriously the rather unfortunate third word in the Birmingham Centre's title – *contemporary*. The irony of course is that the Centre for Contemporary Cultural Studies, the institutional foundations of cultural studies in Britain and beyond, was always concerned with the historical. Its third Director, Richard Johnson, was a trained historian and one of its key publications was *Making Histories* (Centre for Contemporary Cultural Studies 2012). But it is undoubtedly true that cultural studies has not been as historical as its general position on cultural analysis would suggest. It is hoped, therefore, that this volume, by taking the reader into territory too rarely encountered in most versions of the teaching of cultural studies, will bring to the discipline, and to the study of popular culture more generally, a renewed interest in the historical conditions of existence of popular texts and practices.

The new urban society

Between 1770 and 1850, England was transformed from an agrarian to an industrial mode of production, and from a rural to an urban society. The 1851 census revealed that 51 per cent of the population was now living in towns and cities. The impact on popular culture was twofold. There was a change in the content of popular culture, away from the often bloody and brutal practices of the past, to the so-called 'rational recreations' of the present; or, to read it in a very different way, from the communal and self-made activities of the past to commodities and commodified forms of the present. But perhaps more significant was the changed structural position of popular culture. What had once been a shared culture began to split into two, producing quite distinct cultures of social class in which popular culture was now seen as more or less the same as working-class culture (see Burke 2009). Four of the factors that helped produce and support these changes in popular culture are the new social geography; the introduction of the factory system; the development of two often hostile class formations (working class and the middle class); and the emergence of the culture industries.

This was a period of rapid urban and industrial growth. For example, in the 1840s, the number of workers in Britain employed in factories and large-scale industry was more than double the number in the rest of Europe (Foster 1980). Manchester, described by Friedrich Engels as 'the first manufacturing city of the world' (2009: 65), was perhaps the most striking example of the new social geography. In the early eighteenth century, Manchester (including Salford) had a population of about 10,000; by 1841, this had risen to more than 310,000. The city was characterised by increasingly clear divisions between residential areas, working class and middle class. This is not to say that rich and poor had previously lived in close proximity; rather, what I am suggesting is that this is the first time they had so conspicuously lived apart. Writing in 1844, Engels observed:

> by unconscious tacit agreement, as well as with out-spoken conscious determination, the working-people's quarters are sharply separated from the sections of the city reserved for the middle-class… . With the exception of [the] … commercial district, all Manchester proper, all Salford and Hulme, a great part of Pendleton and Chorlton, two-thirds of Ardwick, and single stretches of Cheetham Hill and Broughton are all unmixed working-people's quarters.
>
> (2009: 57–8)

Canon Richard Parkinson, writing two years earlier, also identified the unprecedented scale of Manchester's social segregation:

> There is no town in the world where the distance between the rich and the poor is so great, or the barrier between them so difficult to

> be crossed... . The separation between the different classes, and the consequent ignorance of each other's habits and condition, are far more complete in this place than in any country of the older nations of Europe, or the agricultural parts of our own kingdom.
>
> (quoted in Briggs 1990: 114)

A division introduced by the new relations of the factory system compounded the residential divide. The factory worker knew his or her employer's name because it was told to them or they saw it written across the factory gate. The factory owner, however, did not tend to bother with names; he knew his employees as *hands* – anonymous and replaceable. This was the beginning of a new relationship between employer and employee – what Thomas Carlyle called the 'cash nexus' (quoted in Morris 1979: 22). Gone was any sense of mutual obligation; they were now only connected together for as long as the worker worked, and his or her work was required. Outside this relationship, there was no relationship. As Engels put it: 'The relation of the manufacturer to his operatives has nothing human in it; it is purely economic' (2009: 282). Given such a relationship, it is only to be expected that working and living conditions were often appalling. To be working class in such a place might mean being forced to live 'in wretched, damp, filthy cottages, that the streets which surround them are usually in the most miserable and filthy condition, laid out without the slightest reference to ventilation, with reference solely to the profit secured by the contractor' (75). The full horror of these new social conditions is captured by the low life expectancy rates in Manchester in the 1840s (see Table 1).

After being guided around the working-class areas of Manchester in 1845, an American visitor, Henry Colman, used the language of melodrama to capture the shock and disgust of what he had seen: 'Wretched, defrauded, oppressed, crushed human nature, lying in bleeding fragments all over the face of society. Everyday I live I thank Heaven that I am not a poor man with a family in England' (quoted in Briggs 1990: 116). But as Engels makes clear, the working class were not just victims, as is sometimes their only role in many middle-class accounts; they also resisted and struggled to make something out of their new situation. Therefore, we have to be careful not

Table 1 Low life expectancy rates in Manchester in the 1840s

Classification	*Average age of death (years)*	
	Manchester	*Rutlandshire*
Professional persons and their families	38	52
Tradesmen and their families	20	41
Labourers and their families	17	38

Edwin Chadwick, *Report on the Sanitary Conditions of the Labouring Population of Great Britain*, 1842 (quoted in Briggs 1990: 101)

to see rural and urban workers as the passive victims of these changes. While it is true that throughout the nineteenth century the new urban middle class established hegemony over the new society, we should not think that this means that the working class simply and straightforwardly joined the new consensus. Hegemony is always a continual struggle between incorporation and resistance and as many of the chapters in this collection show, while the middle class was very successful at incorporating the working class, resistance was always a possibility (see Gramsci 1971, 2009, and Storey 2010).

As already alluded to, the third factor is the development of new divisions of social class. Although EP Thompson may be right when he claims that, 'when every caution has been made, the outstanding fact of the period between 1790 and 1830 is the formation of "the working class"' (1980: 212), this has to be seen in the context of the general development of social class in the nineteenth century. As he explains, class is relational:

> Class happens when some men, as a result of common experiences (inherited or shared), feel and articulate the identity of their interests as between themselves, and as against other men whose interests are different from (and usually opposed to) theirs.
>
> (8–9)

Therefore, the making of the working class happened alongside, and was mutually dependent on, the making of the middle class. Moreover, the relationship between these two classes had a powerful impact on the making of popular culture. Many of the chapters gathered in this collection see class and the struggle between these two classes as fundamental to our understanding of the making of popular culture in the nineteenth century.

Of course class was not the only new relationship. The new urban culture enabled and constrained new relations of gender and sexuality. Moreover, everything that was happening so dramatically in England was framed by what was happening in the British Empire, which in turn helped produce new relationships of ethnicity and 'race'. The making of popular culture was not only influenced by these new relationships and their imperial framing, it was also a place where we can witness the playing out of these changes, as can be seen in many of the chapters that follow.

As a relationship of cultural patronage between dominant and subordinate classes began to break down, it was replaced by a new relationship between the emerging culture industries (owned and controlled by the new urban middle class) and the mainly urban working class. Until the advent of postmodernism, dominant definitions of popular culture were a consequence of this new relationship (see Storey 2015). Therefore, the fourth factor determining the new popular culture was the development of a whole range of culture industries, the most obvious being publishing, music hall, the seaside holiday, professional sport and the invention of the 'traditional' Christmas, all of which are discussed in the chapters that follow.

What is popular culture?

Two key concepts of popular culture were developed in the period covered by this book (see Storey 2003). The first was popular culture as folk culture, and the second popular culture as mass culture. Like so much invented in the nineteenth century, these two concepts still have enormous sway over ways of thinking about popular culture.

The first really sustained and detailed intellectual linking of 'popular' and 'culture' was developed in Europe in the late eighteenth century as a result of a growing interest in the culture of the so-called 'folk'. In the late eighteenth century, throughout the nineteenth century and into the early part of the twentieth century, different groups of mostly middle-class intellectuals, working under the different banners of nationalism, Romanticism, folklore and, finally, folksong, 'invented' the first concept of popular culture. For these groups, popular culture is culture that originates from 'the people' (i.e. the 'folk'). This produces a definition of popular culture as something that spontaneously emerges from 'below', something communal and self-made. According to this definition, the term popular culture should be used only to indicate an 'authentic' culture of the people. A significant problem with this approach is the question of who qualifies for inclusion in the category 'the people'. For example, the intellectuals involved in the 'discovery' of the folk distinguished between two versions of the people: the 'rural folk' and the 'urban masses', and only the 'folk' were seen as producers of popular culture.

The 'discovery' of the folk, therefore, not only produced a concept of popular culture as folk culture; it also helped to establish the intellectual tradition of seeing ordinary people as masses, consuming mass culture. This is because the 'discovery' of the rural folk was accompanied, and no doubt driven, by the 'discovery' of the urban masses. If the folk represented a disappearing 'positive' popular, the new urban masses represented an emerging 'negative' popular. As Cecil Sharp, one of the leading figures in the British 'folksong' movement made clear in 1907:

> Flood the streets ... with folk-tunes, and those, who now vulgarise themselves and others by singing coarse music-hall songs, will soon drop them in favour of the equally attractive but far better tunes of the folk. This will make the streets a pleasanter place for those who have sensitive ears, and will do incalculable good in civilising the masses.
>
> (1907: 137)

Sharp is clearly working with two versions of the people and two versions of popular culture. The folk produce popular culture as folk culture; the urban masses consume popular culture as mass culture. In this version of popular culture, it is something imposed by the new culture industries on people who are supposedly passive and easily manipulated, commercially and/or ideologically.

Cultural studies works with a concept of popular culture that is a complex and contextual mix of these definitions. Popular culture is neither an 'authentic' subordinate culture, nor is it a culture imposed by the culture industries. Rather, it is a contradictory mix of both, marked by both forces from both 'below' and 'above', both 'commercial' and 'authentic'. To take Sharp's example of music hall as mass culture, cultural studies would argue that it is 'mass culture' but it is also potentially 'folk culture', in that it expresses themes and ideas that have an authentic existence in the everyday lives of the people consuming the performances. Moreover, music hall songs can be taken from the halls into the everyday lives of these people. In this way, music hall is *made popular* by its audience, which involves a sort of communal making of meaning similar to the workings of a folk culture. But the songs are also deliberately designed to be popular with the audience, to make money and thus suggest the workings of an emerging capitalist culture industry. In this way, then, music hall is a site of both the 'authentic' and the 'commercial', exhibiting features that appear both 'mass' and 'folk'. Ultimately, what it is depends on context (see Storey 2015).

About this book

The book consists of fourteen previously unpublished chapters. Contributors were asked to ensure that the topics covered would involve analysis that increases our understanding of the relations between culture and power. This might be in terms of 'race', ethnicity, class, sexuality, gender or any other marker of social difference. It was also requested that each contain an overview of a particular type of popular culture, make an argument for its significance, and be written in a style that is accessible to undergraduate students studying cultural studies or other related disciplines.

Some of the chapters work with an expanded notion of the nineteenth century, in which it begins in the late eighteenth and ends in the early twentieth century. Chapter 1, for example, argues that the erotic culture of the nineteenth century has its foundations in the eighteenth. The growth of print culture during the eighteenth century extended to the increased number and variety of erotic texts. Through an examination of eighteenth- and nineteenth-century erotic literature, the chapter establishes the tropes and images cultivated in the burgeoning erotic culture of this period. The chapter demonstrates how representations of the male and the female contained within the texts were often used to define, control and contain definitions of sexuality and gender; in particular, there seems to have been a concerted effort to culturally legitimise male power through depictions of sexual authority. The chapter traces the development of the capabilities of erotic culture to propound relations of power both within literature and through the audience's interaction with the texts. It explains how the promotion of these notions of power through dissemination to a wide and diverse (male) audience was made possible by urbanisation and the social spaces engendered by the new urban environment.

Cricket is synonymous with Englishness. Since the nineteenth-century's reinscription of cricket as the gentlemanly game of empire, the sport has remained at the heart of Britain's imperial vision of nationhood. In a speech in 1993, former Conservative Prime Minister John Major rearticulated cricketing's eternality, claiming that even from within Europe 'fifty years on ... Britain will still be the country of long shadows on county [cricket] grounds, warm beer, invincible green suburbs'. Seeking to reassure his audience – the Conservative Europe Group – of the hardiness of Britain's unique characteristics, Major substituted his own cricketing Englishness – associated with the rising middle class and the suburban families of the southeast of England – for Britain. Chapter 2 presents a critical discussion of cricket's advancement as a popular sporting pastime, specifically the radical transformation it underwent to become the crucial moral carrier of nineteenth-century imperial values, and the importance of public schools and literature to this process. It examines how, via the reformation of the public schools, the nineteenth-century middle class ensured that the game became the determining cultural product of empire, proffering the claims to exceptionalism and superiority of Englishness that Britain used to carve out its first world place in the global economy and still uses to maintain its claims to first-placed cultural capital.

The subject of Chapter 3 is music hall, probably the form of entertainment most associated with nineteenth-century England. As it evolved into a highly organised business, the halls often became the site of a struggle between bawdy performers and those who would censor their performances. Anti-music hall campaigner Mrs Ormiston Chant and investigative journalist WT Stead were at the forefront of the demand for censorship. How they articulated their opposition often demonstrated the class divide in perceptions of popular entertainment at this time. Stead, for example, described the halls as 'drivel for dregs'. In doing so, he introduced a key approach in critiques of popular culture – how quickly condemnation of texts or practices becomes condemnation of those who consume them. The chapter also explores what music hall might mean for performers who made a living in the halls. Some could become highly paid stars. Archive sources reveal that female performers found comparative freedom in the halls. Equally, despite the widespread racism in much popular culture of the time, non-white performers could become highly successful. As music hall's popularity slowly declined after the First World War, it was replaced by more tightly controlled music and film industries.

Chapter 4 reveals the problems involved in viewing nineteenth-century photographs as evidence or illustrations of the past. Combining poststructuralist theories of photography and a cultural studies approach to the formation of class and gender, the chapter uncovers the discursive power held in historical photographs. It does this via a historical reading of some of the earliest known commercially made photographs of the English working class. These photographs from the 1860s depict what were colloquially termed 'pit-brow women'; that is, women miners from the pit-head around the Lancashire town of Wigan. The chapter uses these images to illustrate the importance of viewing photographs as historically located forms of communication rather than simply material traces of the past. Paying particular attention to the signs of dirt, gender, coal, spades, backdrops and most significantly fashion (as the women wore trousers), it reveals the role such early commercial photography played in the discursive construction of a specifically working-class definition of femininity. The chapter concludes with a case study which examines how these commercially produced photograph were 'used' in a Parliamentary Select Committee investigation of 1866 as evidence of the 'immorality' of female mine work. To counteract this, mine owners, who wanted the women to continue to work, produced photographs of the same women in their Sunday best, carrying Bibles. The case study demonstrates how gender and class are not essential qualities, but are rather social and cultural constructs inscribed on the body. In this way, the 'truth' of photography is questioned by revealing how the same woman can carry quite different meanings depending on the manipulation of signs – shifting the meaning from dangerous working class/unrespectable to aspiring middle class/respectable.

The nineteenth century witnessed a huge growth in literature aimed at children. However, because these books were relatively expensive, they were beyond the reach of most young readers. Magazines were much less expensive, often selling for as little as a penny an issue. As a consequence, during this period, magazines became a very popular form of reading for children and young adults. Chapter 5 examines the first British magazine to use 'boy' in its title, Samuel Beeton's *The Boy's Own Magazine*. First published in 1855, it claimed a readership of over 40,000 by 1862. But the magazine had to work hard for this success. The chapter shows that an analysis of the magazine's first series (1855 to 1862) reveals fundamental changes in the editor's approach that radically changed the magazine's style and content, and made it into an accessible and, even more importantly, engaging read for 'boys'. A brief comparison with the first volume of the immensely popular *Boy's Own Paper* (1879) shows that the Religious Tract Society had learnt Beeton's lesson well, and continued to develop his strategies for making their magazine popular with boys (and not just their parents).

In order to examine the ways in which the commodity culture of the late nineteenth century developed, Chapter 6 presents a critical overview of three themes in the early history of advertising – one of the key popular cultural forms of the period. First, after a brief introduction to the topic, it considers the repertoire of advertising language which was built up over the period, one in which tropes and modes of address still used in contemporary advertisements were developed. The chapter traces the early history of advertising from simple classifieds to more sophisticated advertisements that drew on an expanded range of symbolic and rhetorical resources. The chapter also examines the development of techniques such as the 'confessional form' and the use of celebrity endorsement as modes of generating interest. Second, the chapter details the ways in which advertising appropriated cultural forms from theatre, news media, visual culture and new forms of entertainment to create a new communicative mode. Finally, the chapter explicates the ways in which advertisers worked to develop a specific 'consumer' subjectivity, one not necessarily constructed from markers of class and gender, but in which racialised tropes played a pivotal role in differentiating acts of consumption. Ethnicity, it is argued, stands in these texts as a cipher for the repressed class narratives, signifying the notion of progress and linking it indelibly with the commodity form. In other words, advertising's modes of representation are rooted in the class and gender conflicts of the nineteenth century.

The industrialisation and urbanisation of the nineteenth century did not only lead to dreadful working and living conditions, it also made holidays possible and desirable for the working class. With the expansion of the railroad system, seaside holidays turned into a truly popular phenomenon, and resorts like Blackpool developed with unprecedented speed and became the symbols of working-class seaside tourism. Chapter 7 analyses the popularisation of the seaside holiday in late-nineteenth-century England, and investigates to what extent the seaside functioned as a place where divisions of social class might be overcome. But the popularisation of the seaside meant not only an increase in numbers: it is also led to a change in holiday-making and the meanings attached to the seaside. Such a cultural transition was not without conflict, as the 'masses' now entered a cultural sphere which had previously been dominated by the middle class. As the chapter shows through an analysis of guidebooks, cartoons and seaside stories, there was a strong awareness of class divisions. Ultimately, the same class divisions that characterised everyday life in nineteenth-century England structured debates about the use of leisure, and even the spatial organisation of the seaside. The seaside holiday thus shows us how central class divisions and social boundaries were at that time, and how impossible and dangerous it appeared to allow the existence of a space where these boundaries might be transgressed.

The nineteenth century saw a considerable increase in fiction written for the urban mass reading market. As Chapter 8 makes clear, aesthetic considerations alone will not do justice to the phenomenon, which critics may shrug off as deficient in plot, style and characterisation. Rather, it is necessary to join literary studies, book history, social and economic history to understand the wide and varied output of popular fictions, ranging from short crime and horror stories, to the immensely successful serialisations of Charles Dickens' novels and the late nineteenth-century Gothic and adventure fiction. In addition to the investigation of the material and economic conditions of popular narrative genres, the chapter critically examines the ideological dimensions of these texts, showing how they provided a platform for the negotiation of some central questions of life in nineteenth-century England. Dominant middle-class discourses, such as those of domesticity and imperialism, were addressed in these popular narratives. What is more, much as they seek to enforce hegemonic middle-class normality, the treatment of these discourses in many cases hints at deep anxieties as to their stability and endurance.

In *The Age of Empire*, Eric Hobsbawm (1987) characterises the period from 1875 to 1914 as the moment of British 'high imperialism'. It was during this period that there is an upsurge in the publication of popular adventure stories aimed mostly at boys. It was through reading these fictions that young readers were introduce to the idea of the British Empire, with its particular views on masculinity, imperialism and British racial superiority. Very often these stories are read in modern editions without illustrations, which can give a very misleading impression of how they might have been read in the nineteenth century. Chapter 9 focuses on a selection of illustrations from three juvenile adventure novels set in India: GA Henty's *With Clive in India; or, The Beginnings of an Empire,* illustrated by Gordon Browne (1884); George Manville Fenn's *Gil the Gunner; or, The Youngest Officer in the East*, illustrated by WH Overend (1892); and Frederick P Gibbon's *The Disputed V.C.: A Tale of the Indian Mutiny*, illustrated by Stanley L Wood (1904). The chapter seeks to show the multiple, frequently subtle and sometimes surprising contributions that illustrations made to a popular genre that reached its apotheosis in the period of British high imperialism.

Chapter 10 introduces readers to the early history of British comics. It examines the context that made their publication possible and focuses on the relationship the comics established with their readership, with class as a key consideration. Looking at examples from the period, the chapter balances consideration of the wider picture with the specifics of individual titles and comic strips to show the basis of their popular appeal. The very first comic, *Ally Sloper's Half-Holiday* (1884), was soon followed by similar publications, such as *Comic Cuts* and *Chips*. These papers were linked by their format and by the appeal of their content to a largely urban working-class readership. Cheaply priced and published weekly, they soon established an enormous audience. *Comic Cuts*, for example, achieved a regular circulation of 300,000. The popularity of these comics took advantage of two significant changes in the latter part of the nineteenth century: the development of cheaper printing technology and the impact of the 1870 Education Act. The latter had led to a dramatic social change whereby many among the industrial working-class had, for the first time, a degree of literacy, certainly enough to read and enjoy comics. A key feature of the comics was the way in which they responded to their core readership and began to shape content to appeal to them. This gave rise to the first generation of British comic book heroes such as Ally Sloper and Mr Chips, iconoclastic characters who began a tradition of irreverence, which was to shape later comic staples such as *The Beano* and *The Dandy*.

The new urban culture was a culture of public speech. From the vast variety of nineteenth-century oral performances, Chapter 11 singles out for special attention the popular science lecture, and more specifically the illustrated astronomical lecture with its fascinating mixture of orality and visuality. Exploring how these ephemeral lecture performances were modelled in newspaper reports and popular fiction, mainly between 1860 and 1900, the chapter also inquires how lectures created authority and legitimacy for the speaker through the spectacle of platform presence, and how lecturer–audience (and audience–audience) interaction was involved with the problem of attention, a vital topic at the root of 'modern' psychological constructions of subjectivity. Non-fictional and fictional (in part, satirical) accounts of popular (astronomical) lectures offer us glimpses of an alternative history, one in which the voice of the lecturing authority becomes a backdrop for the voices of the audience who are directing their attention differently or elsewhere. Popular lectures responded to, and helped to condition, wider changes in English society and culture, and textualised oral performances negotiated pressing nineteenth-century concerns about the nature of authority, class and gender. Popularisers were thus not marginal to scientific research; on the contrary, oral performances of science were part of the cultural process of both asserting and testing the legitimacy of systems of knowledge within the larger and conflicting social and political discourses of the time.

Chapter 12 provides a gendered perspective on early record buying. In the late nineteenth century, shellac discs and associated playback devices were amongst the first mass-produced cultural objects to enjoy the dual status of commodity and personal fetish. They existed at the suture of an increasingly globalised market (the record industry was one of the first to become globalised) and of a domestic, often relatively intimate, realm. The chapter explores the relationship between global, mass-produced cultural products and localised, private acts of musical consumption. Of particular importance is the emergence of late nineteenth-century housewives as the primary record-buying public, and their role as mediators between the market and the home. The chapter also examines the role of independent record dealers (such dealers would usually sell wax cylinders, shellac discs and playback machines, together with bicycles and electrical goods). A complementary exploration of early record catalogues produces an image of early vernacular soundscapes, often characterised by novelties and ephemeral pieces. The focus on musical objects enables Chapter 12 to expose the ambiguous (and enduring) relationship between culture and capital, and the increasing role of cultural commodities in everyday life.

Chapter 13 explores three of the nineteenth century's most enduring cultural inventions: the celebration of the 'traditional' Christmas, association football and folksong. It demonstrates how the role of social class was fundamental to all three in terms of their emergence, but also in how we have come to understand the narrative of the development of these traditions. In the case of Christmas, it was a creation of the new urban middle class, those who had benefitted most from the industrial revolution. A complex mix of fear, celebration and ontological insecurity drove their actions. Moreover, it was an invention that was only indirectly connected to religion and directly connected to industrialisation, urbanisation and the need to secure and sustain hegemony. Association football was also a middle-class creation, or so it is often claimed in many accounts of the history of the development of the game. The chapter challenges this widely accepted assumption by complicating the class relations of the game's invention. Finally, the chapter critically explores the relations of social class that were entangled in the Folk Revival of the late nineteenth century.

It is hard to imagine an account of the popular culture of the late nineteenth century that did not include Jack the Ripper. Interestingly, within the context of late nineteenth-century popular culture, Chapter 14 approaches the Ripper as a fictional rather than historical figure. Adopting three lines of enquiry, the chapter first examines the role of the London press in manufacturing and popularizing the Ripper. It then examines the ways in which the Ripper can be read as part of a tradition of sensational gothic literature, from his roots in penny dreadfuls to his influence on later literary works such as Sherlock Holmes and Dracula. Lastly, there is an examination of the ways in which the Ripper can be read within the context of Whitechapel, becoming a symbol for 'outcast London' (see Stedman Jones 2013). The chapter provides ways of reading the Ripper without becoming tangled in the actual crimes or ever-growing list of suspects and conspiracy theories.

A note on history and the past

If history is a discourse, as it is often claimed, it is always one in relation to materials outside itself. But this relationship is not one in which the materials dictate what can be said or what counts as history. Materials are always mute; it is the work of the writing of history, their articulation in discourse that makes them speak. However, as Pierre Macherey points out:

> The act of knowing is not like listening to a discourse already constituted... It is rather the elaboration of a new discourse, the articulation of a silence. Knowledge is not the discovery or reconstruction of a latent meaning, forgotten or concealed. It is something newly raised up, an addition to the reality from which it begins.
>
> (1978: 6)

In other words, history writing is always an interpretation of the past; that is, we always have to distinguish between 'history' and the 'past'. History does not reproduce the 'objective' truth of the past; it produces a way of understanding what happened in the past. This is not to deny the past existed, nor is it a denial that it still exists in material traces; but what existed is always open to different ways of being understood. It is the writing of history that makes the past historical. This does not mean that history writing is subjective fantasy, but the facts of the past are never enough; they have to be interpreted and made meaningful. Although it is a fact that in the FA Cup Final of 1883 Blackburn Olympic beat Old Etonians 2–1, knowing this tells us very little about the historical significance of the event. A fact it is, but knowing this is only the beginning of historical analysis.

History is never simply about the past; it is always in part about a relationship between the past and the present. Why the FA Cup Final of 1883 is an historical event is the work of the present. The past is full of all kinds of things that may never become historical. The selection of what is historical is made on the basis of a sifting of the past for the significant, and what is seen as significant always depends on our position in the present. Another way of saying this is to acknowledge that the writing of history is always, ultimately, political.

This book is about the past but it is also, and inevitably, an argument about the past, written from a perspective in the present. It may not now be clear how the essays in this collection are written from a perspective in the present, but I am sure that in the future is will be absolutely obvious.

References

Briggs, Asa, 1990. *Victorian Cities*, Harmondsworth: Penguin.

Burke, Peter, 2009. *Popular Culture in Early Modern Europe*, London: Ashgate.

Centre for Contemporary Cultural Studies, 2012. *Making Histories*, Oxford: Oxford Univeristy Press.

Engels, Frederick, 2009. *The Condition of the Working Class in England*, London: Panther.

Foster, John, 1980. 'Introduction' to Mick Jenkins, *The General Strike of 1842*, London: Lawrence and Wishart, pp. 13–19.

Gramsci, Antonio, 1971. *Selections from Prison Notebooks*, London: Lawrence and Wishart.

Gramsci, Antonio, 2009. 'Hegemony, Intellectuals and the State'. In John Storey (ed.), *Cultural Theory and Popular Culture: A Reader*, London: Routledge.

Hobsbawm, Eric, 1987. *The Age of Empire*, London: Abacus.

Macherey, Pierre, 1978. *A Theory of Literary Production*, London: Routledge and Kegan Paul.

Morris, RJ, 1979. *Class and Class Consciousness in the Industrial Revolution 1780-1850*, London: Macmillan.

Sharp, Cecil, 1907. *English Folk-Song: Some Conclusions*, London: Novello.

Stedman Jones, Gareth, 2013. *Outcast London*, London: Verso.

Storey, John, 2003. *Inventing Popular Culture*, Oxford: Blackwell.

Storey, John, 2010. *Culture and Power in Cultural Studies: The Politics of Signification*, Edinburgh: Edinburgh University Press.

Storey, John, 2015. *Cultural Theory and Popular Culture*, London: Routledge.

Thompson, EP, 1980. *The Making of the English Working Class*, Harmondsworth: Penguin.

1 'The man of penetration and the girl of capacity'

Negotiating power in erotic culture

Jenny Skipp

In a letter to the *Spectator,* published in August 1711, a female proprietor complained about the impolite behaviour of the male clientele:

> They strive [to see] who shall say the most immodest things in my hearing. At the same time half a dozen of them loll at the bar staring just in my face, ready to interpret my looks and gestures, according to their own imaginations.

This reflects the sexual frisson of eighteenth-century social spaces and the tendency of men to imagine and configure female sexual response. James Boswell's life attests to the seam of sexuality running through eighteenth-century life, facilitated by the shift to a more urbanised culture and the expansion of sociable environs (Maccubin, 1987). Francis Place's writing confirms this, describing Cock and Hen clubs of his youth, where men and women could enjoy casual sexual encounters as well as singing and drinking (Mason, 1994: 155). Place estimated that there were fifteen such establishments between Blackfriars and Scotland Yard at the turn of the nineteenth century (Clark, 2000: 130). If Walter's story in *My Secret Life* (Anon, 1888) is to be believed, his frequent casual sex in both private and public spaces (see Marcus, 1966) suggests that eroticism continued to flow freely and plentifully through London life in the subsequent century. Such continuity confirms sex and sexuality as an important and integral part of eighteenth- and nineteenth-century life.

As Ian Moulton (2000: 6) has suggested of the early modern period, sexual representation of many different kinds 'permeated literary culture' in a variety of ways. As such, it is possible to investigate the importance of sex to society by an examination of erotic literature. This chapter investigates erotic culture through an examination of obscene, bawdy and sexually descriptive texts produced during this time. Although many may have been, erotic texts were not necessarily supposed to be arousing, but they were intended for the audience to engage with ideas about sexual activities, roles and conduct. It is this, in turn, which invested many of the texts with their erotic function. Many writers protested that their intention was not to inflame lust or excite fantasy, or they claimed that by drawing attention to sexual immorality they hoped to eliminate it. However, these claims were often satirical, and even if they were not, the content was tantalising enough to arouse (at least the interest) of the reader.

There has been much contention about the terminology that should be applied to such material. David Foxon claimed that pornography seems to have 'grown to maturity in a brief period in the middle of the seventeenth century' (1965: ix) while Lynn Hunt assumes that 'pornography' did not constitute a distinct category until the early nineteenth century (1993: 9). However, application of the term pornography presupposes that the work is ultimately concerned with prostitution, and this conceals both thematic diversity and the socio-cultural complexity of this corpus of literature and erotic culture as a whole. However, by broadening the terms of study it is easier to demonstrate how, by strategically creating a 'common' erotic culture, producers managed to codify sexuality and its meanings for a wide audience. As Jonathan Katz (1996: 14) asserts, sexuality was culturally coded for specific purposes, and an investigation that represents the broad basis of erotic culture makes it possible to illustrate how and why certain tropes and images were used to define, control and contain definitions of sexuality. This chapter will show that a common feature of eighteenth- and nineteenth-century erotica was that it propounded relations of power within the text itself, and symbolised an authority between the publication and the (primarily male) reader or audience. Erotic texts, therefore, cohere as a literary genre because of the power relationships they represented or tried to encourage. John Storey (2003: x) reminds us that 'meaning is always the site and the result of struggle', and these exertions are evident in the process of articulating normative sexual practices and gender codes through erotic culture. It is, therefore, imperative to see the widespread nature of discussions and depictions of sex and sexuality and not just those reserved to the most obscene of texts.

The growth of print culture during the eighteenth century extended to the increased number and variety of erotic texts. Significantly, after 1750 the format of erotic literature became more diverse. Alongside the erotic novel and treatise, writers penned short poems, songs, ditties, riddles, jokes, toasts and puns with an explicit content. Consequently, the 1750s and 1760s would seem to represent a time when erotic culture became polarised; when writers aimed at either elite, educated readers or a popular audience by choosing a specific format for their bawdy words. However, in practice this separation was less clear and certain. The first part of this chapter will reflect this and argue that erotic literature was produced and potentially consumed by a socially varied audience. There was an overlap in both the types of erotica that elite and popular audiences had access to and the ways and means of consumption. The second part of this chapter will demonstrate that, regardless of the form and irrespective of the intended audience, the 'message' contained within the texts was often consistent: male sexuality should dominate the female and female sexuality was threatening. Lastly, this chapter intends to establish that much of the basis for nineteenth-century interaction with and understanding of erotic tropes was formulated during the eighteenth century. The process of cultural definition and control of sexuality, particularly female sexuality, begun in eighteenth-century erotic culture, created the impetus for greater control and confinement in the subsequent century.

High or popular erotic culture?

The apparent symbiosis between erotica and developing literary mores, evolving knowledge and intellectual ideas suggests eighteenth-century erotic literature to be a medium which catered to more elite cultural tastes. John Cleland's *Memoirs of a Woman of Pleasure* (1748) and *The History of the Human Heart: or the Adventure of a Young Gentleman* (1749), amongst others, employed novelistic and epistolary techniques. Erotic works which represented or parodied the development of literary, scientific and cultural genres have been well documented, as have the ways authors concealed discussions of sex or descriptions of the body beneath layers of innuendo and metaphor (Porter and Teich, 1994; Peakman, 2003; Harvey, 2004). The botany treatise, for example, was a convenient way to discuss genitalia and the human body, as found in Thomas Stretser's depiction of the penis in *The Natural History of the Arbor Vitae, or, the Tree of Life* (1732). Descriptions and assessments of the landscape, parodying the travel genre, allowed the authorial eye to roam the sexualised body, such as the anonymous *A Short Description of the Roads which lead to the Delightful Country called Merryland* (1742), and writing about science could conveniently cover erotic content, such as James Perry's *Electrical Eel: or, Gymnotus Electricus* (1774), a celebration of the phallus. Erotic texts, therefore, engaged with their cultural and intellectual environment.

However, there are three problems with the discussion of such texts. First, focusing on a pseudo-genre: science, botany, travel, and so on, belies the fact that these texts were usually a celebration of the phallic and male authority, as depicted in the text or as exercised through authorial command over descriptions of female bodies. Second, focusing on 'high culture' publications obscures the real potential power of the erotic medium to disseminate images of gender and sexuality. Third, it should not be supposed that texts which required reader interpretation would be unavailable to more popular audiences. Roy Porter and Lesley Hall (1995) have shown how erotic texts could appeal to readers on different levels. Their study of *Aristotle's Masterpiece,* a midwifery tract first published in Britain in 1690 and continuously published throughout the eighteenth century, illustrates the way readers from the higher orders read the text for pleasure and titillation, while it was used by audiences from the lower sorts to educate and inform themselves about the body and beliefs about sexuality. The key is that, whilst readers with more education may have been more able to critique the narrative offered in the literature, both sets of readers were receiving very similar messages from this text. Such evidence is consistent with a rejection of the 'two-tier' system and focuses on the 'interaction rather than division' of the learned and popular in erotic culture (Burke, 1994: xvi).

Such interaction is apparent from an analysis of erotic texts which represented an interplay between written and oral culture. During the eighteenth century, erotic subject matter was presented in texts that would be considered as more popular culture: poetry, songs, riddles, jokes and toasts, formats which transcribed oral erotic culture. Crucially, these formats were presented in very cheap texts as well as those which were more expensive: for example, *The Covent Garden Jester: or the Man of Fashion's Companion* (1785) and *A New Ballad Inscrib'd to Polly Peachum* (1728) both cost 6d, while *The Merry Companion* (1739) and Captain Morris's songs both cost two shillings or more but contained the same sort of toasts, riddles, prose and songs found in more reasonably priced texts. Many texts advertised their association with oral and performative culture on their title page, such as the anonymous *Shakespeare's Jests: or, the Jubilee Jester: Being a Most curious Collection of Funny Jokes, Merry Stories, Droll Adventures, Frolicksome Tales, Witty Quibbles, Youthful Pranks, Smart Repartees, Double Entendres, Wife-Sayings &c* (1777). Around mid-century, the number of texts, which would have ideally been spoken and shared in sociable environments, increased. *The Puzzle* (1745) was a collection of riddles and conundrums and *The Nuts Crack'd* (1745) the answers. The two separate texts published at the same time suggest that while one person read the riddles, probably to an audience, a second read the answers to conundrums such as 'Why is the Whore like a Comet?' (7). The answer, 'Because she has a fiery Tail' (6). While *The Puzzle* hoped it was 'worthy of so many of our Nobility and Gentry' (i), it was priced at 6d, well within reach of men from the middle and even lower-middle classes. Whilst, as Lisa Sigel (2002) notes, access is not the same as consumption, it does indicate potential interaction. There is also every chance that such texts were bought by a consortium of men, perhaps those bonded together by a club or association. As Jonathan Barry (1994: 103) states, the 'bourgeoisie', including lesser merchants, professionals, shopkeepers and skilled craftsmen, were active organisers and attendees of urban associations. Consequently, the content of such texts could have been widely disseminated. The shift in format correlates with the democratisation of urban social spaces and the nurturing of erotic culture within these environs. Such texts drew on the tradition of bawdy humour which Margaret Spufford (1981) has identified in chapbooks of the seventeenth century. Utilising the oral aspects of erotic culture in the written form was a convenient way to absorb elements of traditional culture and make it consistent with the developing urban environment. It is reflective of Jeremy Black's (2005) assertion that the dichotomy between written culture as progressive and oral culture as conservative is unhelpful. Erotic culture manipulated tradition for self-serving purposes. By building on traditional modes of sharing culture, it could disseminate ideas about gender and sexuality widely; to elites, who Peter Burke (2004: 28) points out, could participate 'biculturally' in both popular and learned culture, as well as to non-elite audiences.

Social spaces and erotica

The performative element of many erotic texts further suggests that the erotic genre was one which was often enjoyed socially and not in isolation, as solely an aid to masturbation, as Tim Hitchcock (1997: 17) suggests. As the quote at the opening of this chapter suggests, the urban landscape facilitated sexual encounters, and social spaces offered an opportunity to discuss sex which, in turn, linked these sites with discourses of sex and sexuality. Matt Houlbrook (2006: 133) has observed that the city and the sexual 'appear culturally and conceptually inseparable'. Indeed, this relationship, fostered during proto-industrial England, would flourish over the course of the next two centuries. Houlbrook (135) continues that elements of industrialisation: size of population and settlements, and understandings of community, leisure time and family life, threatened to erode established social boundaries. Erotic culture attempted to negotiate these tensions by utilising environments, such as the tavern, alehouse and coffeehouse, ideal for reading and sharing erotic literature, to reinvigorate traditional gender roles and rebuild social demarcations. These spaces were used to forge contacts and were sites where disparate groups (a product of a rapidly expanding urban environment) could locate their evolving identity (Barry 1994: 91). As Lynda Nead (2000: 154) has shown, the city brought together heterogeneous groups, and erotic culture enabled a focus to alleviate male heterogeneity. The use of erotic texts in sociable environs offered the opportunity to share 'moral, intellectual and social assumptions' with readers, and to confirm or extend views already held by audiences (Rivers, 1982: 1); in this case, confirming male social and sexual authority.

Both structural and agency elements of erotic culture used sociable spaces to trade, disseminate and appropriately respond to erotic ideas. The tavern, notes Hitchcock (1997: 9), was conducive to expressions of bawdy humour, a tradition established in the previous century. Indeed, taverns, alehouses, coffeehouses and the growth of associational culture were all utilised by and in erotic culture. *The Rambling Fuddle-Caps* (Ward, 1709: 3–9), for example, contained a scene where a patron, invigorated by claret, molests the kitchen wench of a tavern, witnessed by other customers. Such texts built on assumptions that taverns were disreputable and where there was drink there was sex, or at least the allusion to it. The writer of multiple erotic texts, Ned Ward, opened an alehouse near Clerkenwell Green in 1712 and had established himself at the British Coffeehouse near Gray's Inn by 1730. Edmund Curll, author of *Memoirs of a Woman of Pleasure*, held his first auction in the Temple coffeehouse (Straus, 1927: 18). Such evidence testifies to Helen Berry's (2003: 13–16) claim that there was a range of coffeehouses to suit all tastes, and Brian Cowan's (2001) assertion that coffeehouses were not such polite spaces, but could be chaotic and impolite.

It is often difficult to discern the writers and publishers of erotic works. However, while Randolph Trumbach (1993: 253) claims that erotic texts were produced by gentlemen, there were more writers from the middle and lower classes than the elite strata. John Wilkes was the son of a distiller, James Perry of a builder, the prolific Edward Ward was said to be of low parental extraction and Samuel Derrick started writing *Harris's List of Covent Garden Ladies* to free himself from debt. Women cannot be excluded from erotic culture. There is evidence that they wrote, read, purchased and were in the vicinity of performative modes of erotica (Peakman, 2003). Imagining a female audience was a convention of the erotic genre, as was the literary strategy of depicting women perusing material which made them open to sexual advances. Nevertheless, production was decidedly male, and content configured for a male audience.

The erotic cultural elements to male gatherings such as the Medmenham Monks have been well documented (Stevenson, 2001). However, this belies the thread of erotic culture running through many other male gatherings during this period. *The Buck's Merry Companion* of 1766 contained the sub-title *An Entire New and Choice Collection of the Most Excellent, Joyous Songs, Catches and Cantasta, Sung by Several Societies of Bloods, Bucks, Merry Fellows, Choice Spirits and Celebrated Genuisses* [sic] denoting its imagined use by male gatherings. Associational life, which increased through the proliferation of clubs during the eighteenth century, embraced reading and performing (Clark, 1983). By 1800, one third of males claimed membership of the associational culture (Clark, 2000: 1). In 1775, a club in Wych Street in London was dissolved where servants, journeymen and apprentices were charged 4d a meeting. This price included music and provision of 'a female' (130). It is, therefore, important to think about associational culture as being open and inclusive to men from various backgrounds. Furthermore, the clientele of alehouses was also diversifying. Peter Clark (1983: 222–6) has noted that the more traditional customers of the alehouse (labourers, servants, husbandmen, craftsmen, building tradesmen, peddlers, tinkers, soldiers and sailors) were joined by people of good fashion by the 1730s, and by mid-century it was acceptable for parsons and government officials to enter alehouses. Consequently, spaces and situations that were ideal for sharing and responding to erotic culture, and the notions of power and authority contained therein, were available to an increasingly diverse body as the century progressed.

Reading and responding to erotic content in social spaces is consistent with Jeremy Black's (2005: 5) analysis of correspondence from this century which shows that eighteenth-century requirements of politeness were not internalised. Indeed, coffeehouses usually discussed as polite spaces were often no such thing. César de Saussure (1726, in de Saussure 1995: 102), after visiting London, warned that coffeehouses were really brothels where men could be preyed upon by 'dangerous nymphs'. This bawdy atmosphere is reiterated by reports that James Salter's establishment in Chelsea held cabinets of curiosities and rarities for the prurient customer to peruse (Berry, 2003: 15). Mullan and Reid (2000: 17) have pointed out that the eighteenth century was a time when relations between popular and polite were negotiable and constantly redrawn, and this is evident in the erotic undertones being cultivated in 'polite' spaces.

The rise in living standards until the 1790s meant that men with professions such as carpenters, bricklayers, plumbers, masons, plasterers, watchmakers, tobacconists, pewterers and coppersmiths would have been able to afford a multiplicity of erotic texts. Anna Clark (1987: 30) has noted that 'plebeian' men spent money on activities which promoted male solidarity, while Michael Harris (1990: 83–108) documented that sailors in Portsmouth spent sixpence or a shilling on literature, even when they were in financial hardship. Indeed, the combination of an increase in literature which facilitated erotic performance and the cheap price of many erotic texts made them potentially available to a socially diverse audience, and created what Karen Harvey (2004: 39) terms 'a shared erotic culture'. This correlates with Patricia Anderson's (1991: 11) examination of popular culture, as she notes that it was not 'exclusively the experience of any one group or class' but comprised of multiple social layers. The multitude of social spaces offered by increased urbanisation meant that there was a 'plural appropriation' (Chartier, 1987: 6) of the ideas in erotic texts. Erotic culture offered the opportunity for men of different backgrounds to absorb, express and respond to fundamental understandings about sexuality and power constructed in erotic texts.

Erotica and gender ideology

In the 1990s, Tim Hitchcock and Randolph Trumbach used depictions of sex within eighteenth-century culture to discuss gender roles and differences. Both claimed that during this period, heterosexuality became the norm and male homosexuality an unacceptable 'other' which undermined masculinity. However, if Jonathan Katz's supposition that heterosexuality is a modern invention is accepted, then what is present in eighteenth- and early nineteenth-century erotic culture needs to be discussed in terms of potentially procreative sex. It was this 'penil-vagina privilege' (Sigel 2002: 47), and the dominance it offered men, which was a common feature of erotic culture, to which the toast in the title of this chapter 'The man of penetration and girl of capacity' testifies (Anon, *The Buck's Merry Companion*, 1766: 143). Much erotica revolved around the idea that by pursuing gratification and dominance derived through potentially procreative sex men could attain manliness.

Assertion and dominance was often achieved by literal and rhetorical aggression. Elizabeth Foyster (2005) has highlighted the discrepancy between prescribed politeness and actual behaviour during this period by showing that there was increased domestic violence despite the growth in civility (see also Clark 1987). Such violence was often sexualised and focused on breasts and genitals (Foyster 2005: 36). Writers of erotica frequently used aggressive sexual imagery within their work. Aggression was seen as a positive way to gain assertion over women and their bodies. Storming, conquering, possessing, owning and seizing were all common ways to describe sexual activity. Economic and military metaphors both became cultural shorthand for access to authority and masculine ideals (McCormack 2005). The toast 'May our members stand stiff to the commodities of Great Britain' was repeated in three different publications and made a connection between sexual and economic productivity. *The Covent Garden Jester* (Ranger, 1785: 20–1, original emphasis) used seafaring metaphors to connote sexually assertive masculinity and concomitant female passivity:

> Cast off all the tackling about the *breech* ... then take off the apron, and that uncovers the *touch hole.* Next get the *breech* down upon the *bed* then you'll have the *muzzle* high enough for *close work;* and that you know is what English sailors delight in – lastly, take your *rammer* in your hand, *thrust home, discharge,* and that completes the business for the present, till you are ready to *fire again.*

Whilst these actions could be seen negatively, as uncivil and critical of impolite actors, the fact that here, as in other literature, the assertions of aggressive manliness are correlated with Britishness or Englishness suggests otherwise.

Writers presented these prescriptions of masculinity in connection with men of all social origins. Sailors; soldiers; merchants, such as mercers; artisans, such as carpenters and furriers; tradesmen, such as fishmongers, butchers and plumbers, were, amongst many others, featured alongside the gentleman of leisure or disreputable Lord. Frequently, allusions to professions were connected with an assertive, often aggressive male sexuality: 'that the Marrow-bone of thy Maidenhead, being crakt with the Chopping-knife of my courage...' ('Captain Bland', *The Northern Atlantis*, 1713: 245). Writers used erotic culture as an opportunity to disseminate images of manliness and ideal masculinity that mirrored wider efforts to retain male authority over women. In their works, they depicted prescriptive behavioural codes which acted as models of 'normative' masculinity to which their audiences could aspire. It is important to note that Karen Harvey's (2004: 136–9) examination of erotica shows that depictions of the body were counterbalanced by representations of them as soft and vulnerable, whilst Julie Peakman (2003: 92 and 191) suggests that female bodies were not necessarily subservient to male and were, in some texts, depicted as more powerful and sensual. However, these conclusions are based on too narrow a representation of erotic culture and are overwhelmed by the number of texts which demonstrate the opposite.

Features of erotic culture, then, correspond with definitions of patriarchy offered by Heidi Hartman and Eve Kosofsky Sedgewick. Hartman (1997:

101) suggests that there is a material base to relations between men which helps them to create an interdependence and solidarity amongst themselves whilst dominating women. Sedgewick (1985) concurs that the promotion of male self-interests between men is inherent to patriarchy. In erotic culture, there was a forging of fraternity and male privilege through the dissemination, performance and response to erotic literature. The ability of patriarchy not only to adapt to its cultural surroundings but to manipulate them so successfully paved the way for the means of doing so in the nineteenth century.

The operation of patriarchy is also evident in the way writers of erotica approached the representation of female sexuality. All gradations of female sexuality were represented in eighteenth-century erotic culture. While depictions usually fell into two categories – the innocent maid who is deflowered and then receptive, or the libidinous public whore or aristocratic private whore – there were other representations:

> Yet tho' I'm young, I'm not afraid,
> I'll freely f________ like a good natur'd maid;
> There's none in all the town, I'll lay you odds,
> Will sooner tame your P______ or drain your C______.
>
> Anon, *The Gentlemen's Bottle-Companion* (1768, p.57)

However, as above, when the female's sexual appetite was greater than a man's, it threatened masculinity by threatening impotence. Celebration of female sexuality was, therefore, frequently tempered by such fears. As the century progressed, the need to contain the authority sexuality gave to women led to the polarities of the chaste, unknowing but receptive female and the sexually aware, dangerous woman becoming engrained in representations of women in erotic culture. These paradoxes would continue and become amplified in the next century as the ideal of the passive female was strengthened by the depiction of her ideological opposite.

Bradford Mudge (2000: xi) asserts that femininity was used by authors as a 'mirror' to reflect the truth of human sexuality. Lisa Sigel (2002: 2), however, contends that the mirror was shattered, that it reflected, refracted and distorted the picture of sexuality as texts imagined sexual possibilities not necessarily reality. This idea of multiplicity corresponds with Julie Peakman's (2003) findings that it was this multitude of female bodies that leant erotic writing the capability of inverting or subverting dominant ideologies, and this was part of its appeal. However, depictions of sexual women were often hyperbolic or condemnatory:

> As Oyster Nan sat by her tub,
> To show a vicious inclination,
> She gave her noble part a scrub,
> And sigh'd for want of copulation.

A vintner of no little fame,
Who charming red and white can sell ye,
Oserv'd the dirty little dame
As she sat scratching her belly.

Come in says he, you dirty slut,
I've found a rare convenient minute,
I'll ease the itching of your scut,
Or else some greedy devil's in it.

(Anon, *The Gentlemen's Bottle Companion*, 1768: 1)

Importantly, when women were presented as being overtly sexual it was usually accompanied by a rhetoric which disdained their behaviour, showed them to be corrupt or perverted forms of femininity. Descriptions of their sexuality were often coupled with allusions to venereal disease, pride and duplicity.

Writers discussed all types of women to signify the perceived widespread nature of female libidinity. *Nocturnal Revels: Or, the History of King's-Place and Other Modern Nunneries* (Anon, 1779: 238) contained the following empirical estimation of women's sexuality:

1. Furor Uteriuns	2 in 100
2. One digit below the Furor	4 in 100
3. To be completely gratified	6 in 40
4. Extravagant passions	10 in 50
5. Insurmountable desires	12 in 60
6. Enchanting pantings	6 in 20
7. Inordinate titillation	8 in 30

The scale of 'Female Incontinence and Continence' ran to 31 classifications and finished with '31. Cold frigid insensibility – 1 in 100,000', its exaggerations a clear indication of the anxieties female sexuality could create. The author 'Monk' (1779: 270) claimed that by early knowing them, men could come to detest the harlots of society. Labelling female sexuality, therefore, was an attempt to contain it. The author of *the Palace Miscellany* (1732: 67) declared that when women deviated from prescriptive femininity through lustful actions, 'The Bard should then open War declare, And strive, with Satire, to reclaim the Fair'. Writing about sex through discussing femininity was popular because it allowed writers to discuss women's social and sexual roles, usually through a criticism of overtly sexual women, which concomitantly supported male ideals of chaste femininity. As Laura Gowing (2003) notes, female sexuality was inevitable but problematic, and erotic culture offered itself as a medium to deal with this problem by becoming a site where anxieties could be laughed at, female bodies labelled, and sexuality organised in order for men to regain authority.

Writers of erotic literature produced works that suggested, both through rhetoric and content, male power over female figures, and an authority over descriptions of the female body and sexuality. Jennie Mills (2009) has shown that, by distancing men from rape through the formulation of a defloration narrative, writers did not have to worry about the strictures of politeness and sensibility. Female response was ignored, limited or controlled as a necessary way to sanction aggressive male sexuality which became a normal manifestation of gender relationships. The carefully controlled response of women usually depicted them as receptive or submissive in opposition to the assertive male, as the toasts 'may the cushion of life be thump'd by the club of vigour' (Anon, *Shakespeare's Jests,* 1777: 144) and 'May the fair sex suck seed' in *The Buck's Merry Companion* of 1766 (Anon: 143) demonstrate. In turn, male mastery and female submission became culturally coded as erotically appealing (Walker, 1998). Where there was not literal male domination in a text, women's bodies were subjected to the male gaze. The pseudo-topographical text *The Natural History of the Frutex Vulvaria, or Flowering Shrub* ('Clitorides', 1732) defined female genitals, while the pseudo-scientific investigations, *The Semi-Globes or the Electrical Orbs* (Anon, 1777) was a discussion of breasts. This fixation of subjecting the female form and sexuality to descriptive male authority persisted into the next century. Indeed, Walter, in *My Secret Life* (Anon, 1888), continued the trend when he claimed that he had inspected thousands of female genitals (see Marcus, 1966). A problem which both the eighteenth and nineteenth centuries shared was that levels of prostitution, pre-marital and illicit sex meant that neither century could deny female sexuality. This became even more apparent in urban societies where sexuality was more evident. Consequently, by developing specific narratives and images, erotic culture became a means of controlling and containing female sexuality and offered men descriptive and literal power over women.

The nineteenth century

Such labelling became more pronounced as time went on and was fed by the 'preoccupation with order' of the Evangelical movement who made women's 'behaviour, demeanour, social and familial roles, responsibilities and relationships' the focus of their orderliness (Parratt, 2001: 54). Late eighteenth- and nineteenth-century society became more critical of women's participation in leisure time activities. Where once Daniel Defoe had complained of women's impudent, loose and indecent behaviour at the Horn Fair of 1742, the assault on popular recreational culture by spiritual and moral elements, Caitriona Parratt (2001) argues, saw women preferably removed from sites of potential bawdiness (see also Goulby and Purdue 1984: 41–62). Followers of Robert Owen advocated companionate leisure as a means to reconfigure working class gender relations as a prelude to socialism (Parratt 2001: 59). This advocation operated against homosocial gatherings in the

1820s–1840s. However, there was a concomitant movement to associate female respectability with the private sphere and women's withdrawal from the public domain (see Davidoff and Hall, 1987). As a consequence of these attempts at reform and reconfiguration, the women who remained in bawdy atmospheres would have been seen as prostitutes or women inclined to casual sexual encounters, thus confirming the stereotype to producers of early nineteenth-century erotica. As women's participation in popular culture was more carefully controlled, so the fantasy surrounding their response to erotic stimulus increased. The fantasy figure of the inquisitive virgin who becomes overwhelmed by her natural sexual inclinations once deflowered, or even during the process, became a feature of nineteenth-century erotica, as exemplified by *The Ticklish Minstrel* (1837) and *The Lustful Turk* (1828). Such fantasy was a corollary of women's public cultural representation.

Nineteenth-century 'imaginings' (Sigel 2002: 2) of female sexuality led to the female narrative becoming predominant in erotic culture where once it had been shared with the male, libertine narrative. It was increasingly female voices which took on the articulation of male sexual assertion through aggressive actions, as Emily in *The Lustful Turk* (1828) narrates the 'one merciless, violent thrust' her captor uses to deflower her (cited in Marcus, 1966: 215). In *The Festival of the Passions* (1885: 35), Anne watches her friend Lucy and lover George have sex and happily notes that 'after a mutual heave and thrust [George's penis] crammed her to choking'. Female voices articulated the (enormous) size and violence of penises and the liquid deluge that they issued. However, this is not to say that aggressive assertion was not also expressed by male voices or in texts used in homosocial spaces. *Cythera's Hymnal* (1870: 73), for example, contained the following, which it describs as a 'nursery rhyme':

> There was a young lady of Diss
> Who went to the river to Piss;
> The man in the punt
> Shoved the pole up her cunt,
> And gave her the most exquisite bliss.

Violent and aggressive assertions of male writers and characters over female bodies and sexualities were an eighteenth-century tradition embraced and extended by the subsequent century.

Steven Marcus (1966: 230) noted that 'pornography' became much more aware of itself as 'pornography' as the nineteenth century progressed. Indeed, many texts lost their interaction with the literary, scientific and intellectual frameworks enjoyed during the eighteenth century, and became a catalogue of sexual activity which followed formulaic compositions. Common tropes were flagellation, lesbianism, genital inspection, potentially procreative sex and an emphasis on the production of liquids during intercourse, without the detailed development of character and setting that had been prevalent before.

Such formulism may be attributed to the increased professionalisation and commercialisation of erotic culture. For example, by the mid-1830s, there were 57 bookshops dealing in erotic material across London (Sigel 2002: 21). Iain McCalman (1988) attributes this shift to a radical set of cultural producers being drawn into the 'pornography' trade through the Queen Caroline affair, when texts attacked the monarchy and aristocracy through a depiction of their depraved sexuality. The passage of the Blasphemous and Seditious Libel Acts of 1819 forced these producers to abandon their republicanism in favour of the pursuit of 'pornography'. McCalman sees this change in their agenda as antithetical to their political objectives. However, Lisa Sigel (2002) says these producers continued to see 'pornography' as a political medium. Texts contained politicised messages, such as a brand of libertinism that advocated rejecting the strictures of society in favour of a more individualistic and liberal way of living, all expressed through sexual activity.

Sigel (2002: 23) has shown that the radical heritage of many nineteenth-century 'pornographers' and the remnants of their distribution networks allowed erotic wares to be disseminated across class lines as producers continued to target the market that had bought radical texts pre-1819. McCalman (1988) has also noted that William Dugdale's publications cut across working and middle-class boundaries as artisans, clerks, army and navy officers, students, journalists, businessmen and government officials purchased the texts he produced. Just like the eighteenth century, there was a range of prices for erotic texts. *The Exquisite* (1842–4), a weekly publication, was priced at 4d, while *The Modern Rake* (1824) cost a considerable 3 guineas. However, it was the cheap price of *Singular Misadventures of the Right Honourable Filthy Lucre* (n.d.), a text which saw William Stange prosecuted for its publication, which prompted Lord Campbell to enact the 1857 Obscene Publications Act. Sigel has suggested that nineteenth-century erotica was 'a cross-class phenomenon that favoured the well-to-do' (2002: 3). However, while the eighteenth century had established the democratisation of access to erotic culture, the difference a century later was that the acknowledged potential access of the working class precipitated anxiety that led to legislation to limit their potential to engage with such texts.

However, despite attempts to regulate both access and social spaces, erotic culture continued in the nineteenth century, both intentionally and inadvertently. Lynda Nead (2000: 182) has highlighted the fact that Holywell Street, where shops of booksellers who advertised erotic merchandise in their windows were located, was a passage between two main London thoroughfares. As such, men, women and children of all social backgrounds would have been exposed to erotic culture, even if accidentally. Jamie Stoops (2015: 137–56) has shown that while women from the lower classes were purposefully involved in the pornographic trade in the late nineteenth century, the moralistic concerns of the middle classes about the 'poison' of erotic material meant that they were more engaged with erotic culture than was either desirable or officially recognised. However, the retention of literary formats such as songs, poems and toasts, that could potentially be utilised socially, marks the deliberate attempt to use social spaces for erotic culture as a means to disseminate the content as widely as possible. The humour of texts such as *The Ticklish Minstrel* (1837), *Rambler's Flash Songster* (c.1838 and 1865) and *The Gentleman's Spicey Songster* (1841) were dependent on them being used socially, preferably sung. Michael Mason (1994, p.164) notes that the British Library contains over 1,000 individual obscene songs from the period between 1830 and 1840, and amongst these are at least 60 which were sung at the Cole Hole and Cider Cellars. *The Gentleman's Spicey Songster* (1841: 48) includes the toast to 'soldiers [who] demand a good piece, stiff, ram-rod, and a clean touchhole', drawing on the eighteenth-century tradition to include professions of clientele imaginably present in the audience. The same publication also contains a bawdy story featuring a sailor, soldier, lawyer, parson, draper, youth and rake (1841: 42–3), thus catering to a potentially wide audience and one that was imagined to inhabit the social spaces that embraced erotic culture at that time and in the preceding century. This potential social diversity of the audience and their location in sociable situations is confirmed by the process of publishers and sellers of erotica sending out men to country markets, races and fairs to sell their wares (Sigel 2002: 23). While alehouses and public houses became more focused, organised and politicised during the nineteenth century, there would still have been an element of the erotic and they could still be raucous places (Baker, 1998). The Pelican, for example, a late nineteenth-century club, was described as raffish and fast, interested in woman, wine, gambling and song (Lejeune, 1984).

Michael Mason (1994: 163–4) notes that nineteenth-century songs ranged from those containing innuendo to the obscene. Indeed, the tradition of sexual euphemisms remained into the nineteenth century. *The Ticklish Minstrel* (1837: 28–9) euphemistically uses a hat as a vagina, while *The Gentleman's Spicey Songster* (1841) utilises a female tobacconist's trade with her customers as a way to metaphorically refer to sex. Mason (1994: 164) notes that artisans continued to use vocational references to allude to sex in wedding toasts. Although this technique of using innuendos, metaphors and euphemisms had been forged in the eighteenth century to enable participants to circumvent polite prescriptions of society, it may well have gained in prevalence during the nineteenth century as a means of articulating ribaldry whilst still operating within the confines of increased respectability. Such a strategy required knowledge and appropriate response on behalf of the performer and listener, necessitating a shared erotic culture.

Conclusion

Sexuality, and the cultural expression thereof, was frequently a site where anxieties about lost control, changing social structures or gender shifts were played out. Within erotic culture, notions of social and sexual power were negotiated through depictions of appropriate and inappropriate sexual behaviour (Shoemaker, 1998: 67). Both producers and consumers began to codify appropriate and aspirational sexual conduct, dominant and assertive male sexual authority, and idealised passive femininity during the eighteenth century, and this persisted and was exaggerated into the subsequent century. The restriction of (polite) women's access to the sexual elements of popular culture and concomitant cultural representation of them as sober, chaste, diligent and faithful produced a wave of fantasies about their hidden longings and an entrenchment of ideas about publicly sexual women. Such paradoxes of female sexuality were present in the eighteenth century but became more pronounced in the following century as control and containment of female sexuality took a moral and spiritual trajectory. However, men and masculinity were also, in a way, contained and confined during this period, as definitions of assertive manliness through aggressive and potentially procreative sex were enshrined in erotic culture. The 'heterosexual' imperative was consolidated over time, and access to masculinity was formulated through an appreciation and replication of sexually assertive behaviour and attitudes. Male audiences had to respond appropriately to erotic content and participate in cultural formulations of 'normative' sexuality. Such entrenchment had its own confining implications for men as well as for women. Both 'learned' and 'popular' tastes were catered for throughout this period. But it makes little sense to exaggerate this difference as tropes, narratives, rhetoric and imagery suggests the progressive formulation of a 'common' erotic culture which expressed a particular gender consciousness. While the polite may have disengaged themselves from the indecency of popular culture, an erotic culture certainly continued in a multitude of forms in the nineteenth century, and the foundations of this culture were the attitudes to gender roles and an assertion of male power that had been so fundamentally established during the eighteenth century.

References

Anderson, Patricia, 1991. *The Printed Image and the Transformation of Popular Culture, 1790–1860* (Oxford: Clarendon Press).

Anon, 1700. *Aristotle's Master-Piece: or, The Secrets of Generation Displayed in all the Parts Thereof* (London: n.p, original 1694).

Anon, 1732. *The Palace Miscellany* (London: J. Dormer).

Anon, 1739. *The Merry Companion: or, Universal Songster* (London: Printed for Mess, Hazard, Ward and Chandler).

Anon, 1742. *A Short Description of the Roads Which Lead to the Delightful Country Called Merryland. To which are Subjoined an History of the Gallantries of Bettyland. With some Carnal Recreations in Prose and Verse* (London: E. Curll).

Anon, 1745. *The Nuts Crack'd: Being an Answer to the Puzzle: Or a Collection of Conundrums* (London: Printed for M. Cooper).

Anon, 1745. *The Puzzle: Being A Choice Collection of Conundrums* (London: Printed for M. Cooper).

Anon, 1749. *The History of the Human Heart: or the Adventures of a Young Gentleman* (London: Printed for J. Freeman).

Anon, 1766. *The Buck's Merry Companion. An Entire New and Choice Collection of the Most Excellent, Joyous Songs, Catches and Cantatas, Sung by Several Select Societies of Bloods, Bucks, Merry Fellows* (London: Printed for J. Cook).

Anon, 1768. *The Gentleman's Bottle Companion, Containing a Collection of Curious, Uncommon and Humorous Songs: Most of which are Originals* (London: n.p, 1768).

Anon, 1777. *The Semi-Globes, or, The Electrical Orbs. A Poem* (London: Printed for A. Webb).

Anon, 1777. *Shakespeare's Jests: or, the Jubilee Jester: Being a Most Curious Collection of Funny Jokes, Merry Stories, Droll Adventures, Frolicksome Tales, Witty Quibbles, Youthful Pranks, Smart Repartees, Double Entendres, Double Entendres, Wife Sayings &c.* (London: Printed for J. Roson).

Anon, 1779. *Nocturnal Revels: or, the History of King's Place, and other Modern Nunneries. Containing their Mysteries, Devotions and Sacrifices. Compairing also, the Ancient and Present State of Promiscuous Gallantry with the Portraits of the most Celebrated Demireps and Courtezans of this Period* (London: Printed for M Goadby).

Anon, 1828. *The Lustful Turk; or, Lascivious Scenes from a Harem* (London: John Benjamin Brookes).

Anon, 1837. *The Ticklish Minstrel! A Most Bewitching Collection of the very best and most truly Lecherous and Delicious Vocal Tit-bits ever printed, etc.* (London: West).

Anon, 1838. *Rambler's Flash Songster* (London: W. West).

Anon, 1841. *The Gentleman's Spicey Songster* (n.p.: W. West).

Anon, 1870. *Cythera's Hymnal; or, Flakes from the Foreskin. A Collection of Songs, poems, Nursery Rhymes, Quiddities &c. Never before Published* (Oxford [London]: Published at the University Press for the Society for Promoting Useful Knowledge).

Anon, 1885. *The Festival of the Passions; or Voluptuous Miscellany.* By Philo Cunnus (Glenfucket: Abdul Mustapha).

Anon, 1888. *My Secret Life* (Amsterdam: Privately printed for subscribers).

Baker, T. T. ed. 1998. 'Bethnal Green: Social and Cultural Activities' in *A History of the County of Middlesex: Vol. 11, Stepney, Bethnal Green* (London: Victoria County History). Accessed May 10, 2015, http://www.britishhistory.ac.uk/vch/middx/vol11/pp147-155

Barry, Jonathan, 1994. 'Bourgeois Collectivism? Urban Association and the Middling Sort', in *The Middling Sort of People: Culture, Society and Politics in England, 1550–1800*, eds. Jonathan Barry and Christopher Brooks (Basingstoke: Macmillan), pp. 84–112.

Berry, Helen, 2003. *Gender, Society and Print Culture in Late-Stuart England: The Cultural World of the Athenian Mercury* (Aldershot: Ashgate).

Black, Jeremy, 2005. *A Subject for Taste: Culture in Eighteenth-Century England* (London and New York: Hambledon and London).

'Bland, Captain', 1713. *The Northern Atlantis: or, York Spy. Displaying the Secret Intrigues and Adventures of the Yorkshire Gentry: More Particularly the Amours of Melissa* (London: Printed for A. Baldwin).

Burke, Peter, 1994. *Popular Culture in Early Modern Europe, 2nd ed.* (Aldershot: Scolar Press).

Burke, Peter, 2004. *What is Cultural History?* (Cambridge: Polity Press).

Chartier, Roger, 1987. *The Cultural Uses of Print in Early Modern France,* trans. Lydia G. Cochrane (Princeton: Princeton University Press).

Clark, Anna, 1987. *Women's Silence, Men's Violence: Sexual Assault in England 1770–1845* (London: Pandora).

Clark, Peter, 1983. *The English Alehouse: A Social History, 1200–1830* (London: Longman).

Clark, Peter, 2000. *British Clubs and Societies, 1580–1800: The Origins of an Associational World* (Oxford: Clarendon Press).

Cleland, John, 1748. *Memoirs of a Woman of Pleasure* (London: Printed for George Fenton).

Cowan, Brian, 2001. 'What was Masculine about the Public Sphere? Gender and the Coffeehouse Milieu in Post-Restoration England', *History Workshop Journal,* 51, 127–57.

Davidoff, L. and C. Hall, 1987. *Family Fortunes: Men and Women of the English Middle Class 1780–1850* (London: Hutchinson).

de Saussure, César, 1995. *A Foreign View of England in 1725–1729* (London: Caliban).

Foyster, Elizabeth, 2005. *Marital Violence: An English Family History, 1660–1857* (Cambridge: Cambridge University Press).

Foxon, David, 1965. *Libertine Literature in England: 1660–1745* (New York: University Books).

Goulby, J.M. and A.W. Purdue, 1984. *The Civilisation of the Crowd: Popular Culture in England 1750–1900* (London: Batsford Academic and Educational).

Gowing, Laura, 2003. *Common Bodies: Women, Touch and Power in Seventeenth-Century England* (New Haven and London: University of Yale Press).

Harris, Michael, 1990. 'A Few Shillings for Small Books: the Experience of a Flying Stationer in the Eighteenth Century,' in *Spreading the Word: the Distribution Networks of Print, 1550–1850,* eds. Robin Myers and Michael Harris (Winchester: St. Paul's Bibliographies), pp. 83–108.

Hartman, Heidi, 1997. 'The Unhappy Marriage of Marxism and Feminism: Towards a More Progressive Union', from *The Second Wave* edited by Linda Nicholson, (London: Routledge,) pp. 97–123.

Harvey, Karen, 2004. *Reading Sex in the Eighteenth Century: Bodies and Gender in English Erotic Culture* (Cambridge: Cambridge University Press).

Hitchcock, Tim, 1997. *English Sexualities, 1700–1800* (London: Macmillan).

Houlbrook, Matt, 2006. 'Cities' in Matt Houlbrook and Harry Cocks eds. *The Modern History of Sexuality* (Basingstoke: Palgrave), pp. 133–57.

Hunt, Lynn, 1993. *The Invention of Pornography: Obscenity and the Origins of Modernity, 1500–1800* (New York: Zone Books).

Katz, Jonathan Ned, 1996. *The Invention of Heterosexuality* (New York: Plume).

Lejeune, Anthony, 1984. *The Gentlemen's Clubs of London* (London: MacDonald's and Janes).

McCalman, Iain, 1988. *Radical Underworld: Prophets, Revolutionaries and Pornographers in London, 1795–1840* (Oxford: Clarendon Press).

McCormack, Matthew, 2005. *The Independent Man: Citizenship and Gender Politics in Georgian England* (Manchester: Manchester University Press).

Maccubbin, Robert P., 1987. *Tis Nature's Fault: Unauthorized Sexuality During the Enlightenment* (Cambridge: Press Syndicate of the University of Cambridge).

Marcus, Steven, 1966. *The Other Victorians, A Study of Sexuality and Pornography in Mid-Nineteenth-Century England* (New York: Basic Books).

Mason, Michael, 1994. *The Making of Victorian Sexuality: Sexual Behaviour and its Understanding* (Oxford: Oxford University Press).

Mills, Jennie, 2009. 'Rape in Early Eighteenth-Century London: A Perversion "so very Perplex'd"', in Julie Peakman ed. *Sexual Perversions, 1670–1890* (Basingstoke: Palgrave Macmillan).

Moulton, Ian Frederick, 2000. *Before Pornography: Erotic Writings in Early Modern England* (Oxford: Oxford University Press).

Mudge, Bradford K., 2000. *The Whore's Story: Women, Pornography and the British Novel, 1684–1830* (Oxford: Oxford University Press).

Mullan, John and Christopher Reid, 2000. *Eighteenth-Century Popular Culture: A Selection* (Oxford: Oxford University Press).

Nead, Lynda, 2000. *Victorian Babylon: People, Streets, and Images in Nineteenth-Century London* (New Haven, CT: Yale University Press).

Parratt, Caitriona, 2001. *More than Mere Amusement: Working Class Women's Leisure in England, 1750–1914* (New England: Northeastern University Press).

Peakman, Julie, 2003. *Mighty Lewd Books: The Development of Pornography in the Eighteenth Century* (Basingstoke: Palgrave).

Porter, Roy and Lesley Hall, 1995. *The Facts of Life: Creation of Sexual Knowledge in Britain. 1650–1950* (New Haven, CT and London: Yale University Press).

Porter, Roy and Mikuáš Teich, 1994. *Sexual Knowledge, Sexual Science: The History of Attitudes to Sexuality* (Cambridge: Cambridge University Press).

Ranger, Roger, 1785. *The Covent Garden Jester, or the Rambler's Companion* (London: J. Walker).

Rivers, Isabelle, 1982. *Books and their Readers* (Leicester: Leicester University Press).

Sedgewick, Eve Kosofsky, 1985. *Between Men: English Literature and Male Homosocial Desire* (New York: Columbia University Press).

Shoemaker, Robert B., 1998. *Gender in English Society, 1650–1850: The Emergence of Separate Spheres?* (London: Longman).

Sigel, Lisa Z., 2002. *Governing Pleasures: Pornography and Social Change in England, 1815–1914* (New Brunswick, NJ: Rutgers UP).

Spufford, Margaret, 1981. *Small Books and Pleasant Histories: Popular Fiction and its Readership in Seventeenth-Century England* (London: Methuen).

Stevenson, David, 2001. *The Beggar's Benison: Sex Clubs of Enlightenment Scotland and their Rules* (East Lothian: Tuckwell).

Stoops, Jamie, 2015. 'Class and Gender Dynamics of the Pornography Trade in Late Nineteenth-Century Britain', *The Historical Journal,* 58: 1, pp.137–56.

Storey, John, 2003. *Inventing Popular Culture* (Oxford: Blackwell).

Straus, Ralph, 1927. *The Unspeakable Curll* (London: Chapman and Hall).

Stretser (Stretzer), Thomas, 1732. *The Natural History of the Arbor Vitae, Or, the Tree of Life* (London: Printed for John Wilkinson).

Trumbach, Randolph, 1993. 'Erotic Fantasy and Male Libertinism in Enlightenment England', in *The Invention of Pornography: Obscenity and the Origins of Modernity,* Lynn Hunt ed. (New York: Zone Books).

Walker, Garthine, 1998. 'Reading Rape and Sexual Violence in Early Modern England', *Gender & History*, 10: 1, pp. 1–25.

Ward, Ned, 1709. *The Rambling Fuddle-Caps: or a Tavern Struggle for a Kiss* (London: Printed and to be sold by H. Hills).

2 'But it's more than a game. It's an institution.'

Cricket, class and Victorian Britain's imperial Englishness

Claire Westall

The telling of cricket's past

Derek Birley was right when he explained in *The Willow Wand* that 'cricket is not so much a game as institutionalised nostalgia' (2000 [1979], p. 8). Cricket is typically seen as having a notably long history and is quite clearly besotted with the telling of its past. The game's origins are unknown but a plethora of theories about its name and lineage circulate. Claims to cricket's ancient pedigree often cite an expenses entry of Prince Edward's in 1300 incurred for playing 'creag' (Birley, 1999: 1). They also link 'cricket' to French, Norman, Dutch and Anglo-Saxon terms, and suggest that 'club-ball' and 'stoolball' may be cricketing antecedents. The Anglo-Saxon word 'cricce' or 'cryce' is repeatedly invoked as referencing a staff used by shepherds and has helped breed one of the game's most enduring 'fantasies of the romantic imagination' – that of shepherds using their crooks to play an early version of cricket (Bowen, 1970: 32–35; also see Bateman, 2011: 10–11).

Cricket did not appear in James I's 1618 Declaration of Sports, implying that it was not popular, fashionable nor disorderly enough to warrant a Sunday prohibition. However, the game was enjoyed by schoolboys during the Interregnum, and became popular with Charles II and a fashionable set around court during the Restoration (Birley, 1999: 10–11; 15). Retaining its rural 'plebeian image' (15), though, cricket was played with local variations and customs across the south, particularly in the southeast, and increasingly in the north of England as the eighteenth century advanced. It was, by and large, a two-stump, double-wicket, under-arm game (with some single-wicket variants) that came into a period of rapid regulatory development at the close of the century. In contrast to its popular image as a pure and tranquil rural undertaking, cricket was one of the Regency sports associated with unruly and boisterous crowds, notable drinking and the growing 'national addiction' to gambling (16). Moreover, it was the game's ties to the landed aristocracy and their need to settle sizeable wagers that drove cricket's modern standardisation. In 1727, the first Articles of Agreement were drawn up in anticipation of two (heavily wagered) matches arranged by Charles Lennox (2nd Duke of Richmond) and Mr Alan Brodrick (heir to Viscount Midleton) (18–19). By 1744, Lennox was helping establish the first 'Code' of the game, a code revised in 1755 and again in 1774. Yet it was not until the 1770s that the three-stump wicket came into use. By the 1780s George Finch (9th Earl of Winchilsea) had established the White Conduit Club in Islington and, with Colonel Charles Lennox (4th Duke of Richmond), commissioned Thomas Lord to acquire land for a new private cricket ground. This became Lord's Cricket Ground and lead to the formation of the Marylebone Cricket Club (MCC) in 1787. Importantly, as Mike Marqusee says, cricket was thereby in the hands of those London-linked men who led the world's 'first market economy' and 'took for granted their right to rule at home and abroad' (1994: 53).

In his 1994 book, *Anyone But England*, Marqusee, the American-born cricketing Marxist, unpacks the history and late-twentieth century malaise of English cricket. He describes the game as an aestheticised violent encounter between batsman and bowler that hinges on a critical tension between 'no determined end' and the exciting potential for 'end at any moment' (21). Marqusee understands cricket as premised upon an ideal of continuity that is best captured in an unbeaten innings or a drawn (or tied) match but is often imagined through the valorisation of longevity – of a game, of a day's play, or of a team's or batman's innings. He also recognises that this is bound to the omnipresent threat of rupture, of an irreversible end – as in the fall of a wicket. Importantly, Marqusee maps this tension onto the game's history, asserting that the 'need to prove continuity – direct descent from old England – has been a compulsion for nearly all the game's historians' (47). Further, he contends that a kind of continuity narrative – invested with the lies of Englishness built into cricketing myth – runs directly against the radical transformation the game underwent in the late eighteenth century (46–67), and, I would add, again in the mid-nineteenth. He positions the game's Englishness as 'the lie' of 'the cult of the honest yeoman and the village green' as well as the denial of the game's modern ties to 'commerce, politics, patronage and an urban society' – all of which meant it 'had to be wrapped in something finer and, at least in appearance, older' (71). This ideological wrapping – of cricket as an ever-lasting rural folk game testifying to English sanctity, purity and stability – was cemented during the nineteenth century by two specific cricketing histories.

On the back of an industrial England's nostalgia for the lost village fuelled by William Corbett and Mary Mitford in the 1820s (Bateman, 2011: 12), *The Cricketers of My Time* was published in 1832 in serial form, before becoming part of *The Young Cricketer's Tutor* in 1833. Written with Charles Cowden Clarke, a Shakespearian scholar, this work inscribed Hambledon, of Hampshire, and its Broadhalfpenny Down pitch as the 'cradle' of cricket. From the late 1750s through to the 1780s, Richard Nyren, John's father, was the landlord at the Bat and Ball Inn, organising cricket matches, in which he would regularly play, and an attached social club. Thanks to his son's memorialisation of the club and its innovative players, Hambledon became the quintessential village referent for cricket's eighteenth-century development, and its mixing of classes, despite the ways in which matches were organised by recruiting local professionals to play on behalf of interested gentlemen and imbued with the drinking and gambling of the age (see Birley, 1999; Bowen, 1970; Green, 1988). As Benny Green explains, Hambledon was the first cricketing club to inspire literary work, but the chief Hambledon figures were 'men of London' and so the rural location of literary inspiration should not detract from the wealth-based connections between country and city that were being refined (1988: 13; 25). Anthony Bateman rightly views Nyren's work a 'deeply nostalgic ... foundation-stone of English cricket's literary canon', and as 'part of an emerging literary and artistic sensibility that was crucial in defining a new conception of manly beauty' (2009: 24–25). He also pinpoints how cricket was being 'identified with its past' – imagined more than real – 'as it was inaugurated as a product of modernity' (2009: 23). By mid-century, Reverend James Pycroft's *The Cricket Field* (1851), usually labelled the first full history of the sport, was balancing an awareness of gambling and match fixing with an insistence on cricket's iconic rural Englishness and character-forming attributes. To Pycroft, cricket 'is a game for Britons generally', 'a standing panegyric on the English character' and evidence of an 'orderly and sensible race' possessing '[p]atience, fortitude and self-denial', as well as 'an unruffled temper' and 'good generalship' (1854: 16; also cited by Bateman, 2009: 1). Although Pycroft contrasts the English with Irishmen and Scots, the slide from 'English character' to 'Britons' is a common move in depictions of cricketing Englishness and indicative of Britain's imperial vision of Englishness – a vision that overrides England's status as a nation. Notably, these histories were creating a cricketing past of domestic Englishness at the highpoint of the game's middle-class transformation and Britain's supposedly moral claims for imperial expansion.

Victorian cricket's imperial Englishness and public school ideology

Cricket was being played across England at the start of the nineteenth century, and was part of a mass shift towards sporting leisure that accelerated as the century unfolded. With the population of Britain doubling between 1800 and 1851, the year of the Great Exhibition (Golby and Purdue, 1999: 88), the large influx of people into major industrial towns and cities increased sporting participation and spectatorship. According to Wray Vamplew, industrialisation influenced sport in England in three overlapping ways: first, by changing the patterns of work and leisure, thereby separating sporting activities from the agricultural calendar; second, through the patronising of works teams that industrialists undertook to support sport's 'utilitarian function'; and, finally, with 'increased real incomes and leisure time ... stimulat[ing] a demand for commercialized spectator sport' (1988: 17). Football, rugby, cricket and other sporting clubs sprung up all over the country, with the peak for new sporting organisations coming between the 1860s and 1880s. Christopher Brookes sees cricket as having a particular role in limiting the 'impact of [its] new working-class presence' through the creation and maintenance of a 'pair of roles' – amateur and professional – 'one of which affirmed the rights and privileges of the gentleman while the other defined the duties and obligations of the artisan', or worker. This, Brookes argues, allowed cricket to become popular 'without reneging on its elitist traditions' (1978: 138). The corresponding Gentleman versus Players match began in 1806, around the time of the first Eton versus Harrow and Oxford versus Cambridge matches, but it really took hold after the Napoleonic wars, becoming an annual fixture from 1819 until its demise in 1962. While professionals and amateurs existed up and down the country, the north of England became associated with hard-playing league professionals, and the south with amateur-led county cricket. The professional was thought of as the grafting fast bowler tainted by his association with labour, on and off the field, whereas the amateur was the elegant gentleman batsman whose performances were measured in aesthetic and moral rather than economic terms. (This division would also be racialised in empire.) By the Victorian period, Marqusee states, cricket was 'unique among gate-money sports ... in encompassing amateur and professional in one competitive structure' (1994: 96).

For C. L. R. James, if cricket was formed during the lifetime of William Hazlitt (1778–1830), it was under the reign of the 'creators of Victorianism', namely the three public school Thomases – Arnold, Hughes and Brown – and W. G. Grace that the game came to dominate the popular imagination (2000: 160). The ascension of Queen Victoria in 1837 coincided with the 50th anniversary of the MCC, the death of John Nyren and the beginning of a major development in cricketing technique as bowling evolved from a round-arm practice to an over-arm form of attack, roughly from 1835 to 1860 (see Bowen, 1970: 100–105). Victorianism, according to Boyd and McWilliam, though, arrived somewhat belatedly, in the 1840s, coming with rising living standards and expanding middle-class cultural habits (2007: 3). By this point, William Clarke had developed a touring All England XI that was an explicitly commercial venture and part of W.G. Grace's childhood memories; Charles Dickens had penned *The Pickwick Papers* (1836–7) with its Kentish Dingley Dell cricket team; and the Commander-in-Chief of the army had declared that a cricket pitch should be near to each and every barracks (Green, 1988: 41). In the wake of the 1832 Reform Act and subsequent Chartist demands for democratic advancements, cricket underwent its most fundamental ideological rebranding, becoming the sport of the middle class and their reformation of the most significant of British institutions – public schools. Public schools had been spaces of unruly aristocratic excess, of the brutalities and indulgences that had come to characterise the aristocracy and landed gentry (linking them to the working classes). By 1861, the Clarendon Commission had been set up to examine their lack of financial, administrative and educational direction. Reporting in 1864 and underpinning the 1868 Public School Act, the commission's demands for widespread reorganisation, regulation and educational management – including curriculum management – concretised the upper-middle-class hold on public schools, a hold expressed in a new and fundamentally imperial cricketing ideology.

In his authoritative body of work, sports historian J. A. Mangan has established that from the middle of the nineteenth century, public schools functioned as imperial training grounds for Britain's middle-class boys (see 1981 and 1988b). 'The public schoolboy learnt *inter alia* the basic tools of imperial command: courage, endurance, control and self-control', and principally did so on the sports field (Mangan, 1988b: 18). This system of instilling manliness simultaneously taught dominance and deference, agency and subjection, 'initiative and self-reliance but also loyalty and obedience', making it an ideal 'instrument of colonial purpose' (ibid). After a period of 'emergent athleticism' (1800–60) and an 'age of transition' (1860–80), it was during the 1880s that organised games secured a primary place within the public school system (see Chandler, 1991). Thereafter, a Cult of Athleticism flourished that was characterised by prestigious school and inter-school matches, the colours system, sporting masters, House and school loyalty, and the schoolboy sporting hero (Richards, 1988: 120). This hero increasingly embodied the nineteenth-century move toward Muscular Christianity as espoused most famously by Charles Kingsley and Thomas Hughes. The diffusion of these ideals occurred through a process of 'circular causality', with public schools leading Oxbridge only to then receive games-orientated Masters back from the universities, or for such men to take up posts within the empire (see Chandler, 1991). At the heart of this dominant cultural milieu were the three Cs – Classics, Christianity and Cricket. And from the 1850s on, cricket was becoming the vehicle and measure for all that was 'good' and 'true', 'straight' and 'fair'. The phrases 'a straight bat' and 'it's not cricket' came to stand for an ideology of fair play on, off and beyond the cricket pitch. This fair play ethic functioned as a disciplinary tool to control schoolboys and, from there, the imperial nation meeting, as Mangan records, a practical need for social cohesion. It was also 'quickly equated with moral "manliness" – the robust manliness of imperialists' (Mangan, 1996: 34). Jack Williams has highlighted that the sportsmanship of cricket expressed in fair play 'was seen to produce a distinctly national character' (1999: 2) – where character is a euphemism for a masculinity that is inherently (white) British (imagined as always English), athletic and suited to imperial rule. This was the cricketing hero of the public school who was, as an ideal type, represented in, if not created by, public school literature, but who was also sometimes problematised within its pages.

Literature and cricket are so tightly interwoven that, according to Green, 'it is almost as though the game itself would not exist at all until written about' (1988: 14). It would certainly have not become Britain's bearer of imperial Englishness without public school literature, and especially Thomas Hughes' defining work *Tom Brown's School Days* (1857). Running to 11 editions in its first year (Golby and Purdue, 1999: 162), Hughes' popular novel grew to exert a genuinely global influence as *the* text of imperial boyhood. Jeffrey Richards argues that it is 'a mirror of the changing nature and structure of Victorian values, a vital point of intersection between the ideas that had been dominant and the ideas that were to become so' (1988: 24). Using his memories of Rugby School under the headship of Thomas Arnold, from 1834 to Arnold's death in 1842, Hughes connects his vision of educational reform and what becomes Muscular Christianity with an advocacy of Anglo-Saxon English chivalry and ruralism which together form the basis of Britain's imperial supremacy. Indeed, Tom is a 'middling' boy with the 'commonest' of middle-class names (Hughes, 1861: ii) that marks a fighting 'clan' to whom Britain owes much of her 'greatness' (1989: 1). The Browns are 'scattered over the whole empire' (5) and, in Tom's case, settled in the Vale of the White Horse, Berkshire, that 'sacred ground for Englishmen from Alfred onward' (10), and 'didn't go out of the country once in five years' (17). Consequently, the Browns are simultaneously at home and abroad, the essence of England's green locality and Anglo-Saxon past and the foundations of Britain's imperial conquest from America to India (with the book being published in the year of the Indian Mutiny).

In spite of his concern for the Arnoldian prioritisation of religious and moral principles over gentlemanly conduct and intellectual ability, Hughes' hero prioritises becoming 'A1 at cricket and other games' over all else (1989: 255). Hughes' work thereby fuels a move away from Arnold's policy of 'Godliness and Good Learning' (see Newsome, 1961), a move reinforced by the emergence of Social Darwinism and the later dominance of athleticism. By signing the text 'An Old Boy', Hughes also registered the network of former public school boys that was consolidating in the 1850s and, in cricketing terms, was represented by I Zingari (founded by ex-Harrovians in 1845). Muscular Christians were the men Hughes was aiming for, and which he promoted in *Tom Brown at Oxford* (1861), but instead sporting boys, particularly cricketing boys, became the imperial men of Britain's future. Tom is an exemplary case. His heroic development is traced through different sporting encounters, first in the Vale and then at Rugby. His early experiences of country sports come after he has escaped the female/feminising domestic sphere into a world of Regency masculinity featuring backswording and wrestling. Thereafter, Hughes charts the transition from Regency masculinity to Christian gentlemanliness via Tom's school sporting journey at Rugby, from football through boxing to cricketing leadership. The climax of the novel sees Tom, the new 'Nestor of the school' (361), captain Rugby against the MCC. It is hoped that Tom is 'as much wiser as he is bigger' (351), where his physical maturity harps back to a Regency pride in size and strength and stands as evidence of his newly found athleticism which, it seems, is bound to his popularity, leadership and sense of communal responsibility.

In the novel's iconic cricketing scene, Tom converses with a Young Master about the significance of the game as Arthur, his best friend, sits at their feet in a posture of admiration for the two halves that constitute his own character – scholarly Greek/Classics/religious instruction and maturing athleticism/cricket (which rescued Arthur from effeminate isolation). The Young Master is a fictionalised rendering of G. E. L. Cotton (1813–66), a pupil of Arnold's who worked as an assistant at Rugby between 1837 and 1852. Cotton moved to Marlborough to be Headmaster after the so-called last Great Rebellion of 1851, and then to India to become Bishop of Calcutta in 1858. Though Cotton was an intellectual of Arnold's circle, he took inspiration from Charles Vaughan, the Head at Harrow between 1844 and 1859, and deployed competitive sports to rein the school in after a past of 'poaching, trespassing and general lawlessness' (Mangan, 1981: 22), where lawlessness may carry a euphemistic tinge. Thus, his cricketing conversion in *Tom Brown* is a prelude to the change that occurred around the 1860s from boys taking the lead in games, to games being organised by masters as disciplinary mechanisms. It also marked the emergence of a new ideological importance for cricket as a sanctified British institution:

> 'What a noble game it is, too!' [The Young Master says.]
>
> 'Isn't it? But it's more than a game. It's an institution,' said Tom.
>
> 'Yes', said Arthur, 'the birthright of British boys old and young, as habeas corpus and trial by jury are of British men.'
>
> ...
>
> 'The discipline and reliance on one another which it teaches is so valuable, I think,' went on the master, 'it ought to be such an unselfish game. It merges the individual in the eleven; he doesn't play that he may win, but that his side may.'
>
> ...
>
> 'Tom, you've a great deal to learn yet in the art of ruling.' [Arthur]
>
> 'Well, I wish you'd tell the Doctor so then, and get him to let me stop till I'm twenty. I don't want to leave, I'm sure.'
>
> 'What a sight it is', broke in the master, 'the Doctor as a ruler! Perhaps ours is the only little corner of the British Empire which is thoroughly, wisely and strongly ruled just now. I'm more and more thankful every day of my life that I came here to be under him.'
>
> (Hughes, 1989 [1857], 354–356)

Cricket's claim to ethical nobility and imperial preparation is explicit in 'Tom Brown's Last Match' and much has been written about its importance to cricket's popular image. In *The Imperial Prospects of Education 1894–5*, J. E. C. Welldon, Headmaster of Harrow (1885–98) and also a Bishop of Calcutta (1898–1902), almost directly repeated the content of Hughes' cricketing exchange in order to establish how '[i]n the history of the British Empire it is written that England owed her sovereignty to her sports' (cited in Mangan, 1988b: 36). Famously, Henry Newbolt's 1897 poem 'Vitaï Lampada' ('The Torch of Life') provides a paradigmatic example of the coming together of Englishness and empire in the cricketing hero. Its haunting poetic refrain 'Play up! Play up! and play the game!' became an indoctrinatory banner insisting on heroic self-sacrifice and glorious camaraderie in battle as in cricket that held sway until World War I. Patrick Howarth (1973) and Isabel Quigly (1984) agree that Tom Brown functioned as a character mould whose literary life only came into crisis after World War II. Howarth makes clear that this figure merged with Newbolt's, becoming the *homo Newboltiensis* or Newbolt Man – a real as well as literary figure – 'imbued with a strong sense of institutional loyalty, upper middle class by background' and enjoying a 'natural moral prerogative' because of his nation's right to rule, who 'accept[s] ungrudgingly the demands of service and duty' and demonstrates 'natural power of command, some degree of worldly success, a distrust of latter-day politicians and a tendency toward philistinism in artistic taste (1973: 14).

Less attention, though, has been granted to the details of the institutional claim for cricket. Arthur's comparison insists that cricket is the training ground for British manhood, but it also links cricket and the law, establishing a parallel between the individual boy playing within a sporting collective and the state's protection of individual and communal rights to freedom. This parallel encodes the individual's subservience to the team, where that union is Britain – rather than England – and the individual is forever indebted to – rather than being the creator of – the state for the liberties imagined to be both natural, as birthrights, and constructed by the institutions – legal and political – of the state. In this way, cricket is held as a microcosm of British constitutionality where its 'laws', its rules and regulations, sit within the 'spirit of the game' – an extra juridical aura of ethical superiority and uncodified but always-already known conduct – often understood as fantastically English yet bound to the eccentricities and complexities of some ahistorical set of English characteristics that ethically underpin Britain's supposed right to empire. We should read Neville Cardus' later claim for cricket similarly. In his book *Cricket* Cardus wrote that:

> If everything else in this nation of ours were lost but cricket – her constitution and the laws of England of Lord Halsbury – it would be possible to reconstruct from the theory and the practice of cricket all

> the eternal Englishness which has gone to the establishment of that Constitution and the laws aforementioned'.
>
> (Cardus 1931, cited in Birley, 2000: 13)

Birley notes the Tom Brown link but dismisses Cardus' view entirely (2000: 13). However, Cardus' assertion makes explicit the naturalised claims of cricket and their similarity to those of the British state, even as the state masks itself in the glories of imperial Englishness. In Cardus' statement, the 'nation' appears to refer to England despite the post-union constitution at issue being Britain's, or the UK's, and confusion over the 'nation' described is exacerbated by the reference to Lord Halsbury's Laws of England, first published as 31 volumes between 1907 and 1917, and covering England and Wales in an enormous project of alphabetising the laws of the land.

In addition, Arthur positions cricket as indicative of the Englishness at the core of the British Empire. This is reinforced by his sense of the Doctor; that is, Arnold, as an omnipresent 'ruler' of a domestic imperial space. In this move, cricket marks an Englishness bound to British law and order, meaning that discursively the exchange moves from cricket through England to Britain's empire, from the inside out, or from empire through England to cricket from the outside in, though both England and empire are also simultaneously held within the space of imperial cricket. This is an example of the imperial dialectic of Englishness, wherein, as Ian Baucom explains, 'Britain' and 'British' is elastically used to claim the empire and its subjects, while also acting as a buffer between England and empire so that empire is 'within the boundaries of Britishness and outside the territory of Englishness' (1999: 8) – an ethically superior Englishness which must be admired and copied despite its impenetrable exclusivity. Baucom reads the cricket field – alongside other Victorian sites of empire – as an auratic, identity-inscribing locale, a *lieu de mémoire* in Pierre Nora's terminology, whose metonymic qualities create a degree of sameness that, because of multiple sites of transference to, or recreation within, empire means that it is a field of instability for Englishness (1999:18). In this fashion, the Rugby cricket pitch is linked to and established by the other pitches of empire. Baucom also describes Arnold's tenure as Head at Rugby as 'an event in the discourse of Englishness' and endorses Raymond Williams' argument in *The Long Revolution* (1961) that Arnold oversaw 'the arrival of a middle class dominance in the control of public schools and precipitated the rapid and massive expansion of those schools and their ability to discipline a hegemonic representation of English cultural identity' (138) – where 'English' stood for Britain's version of an imperially oriented and bound ideological construction, and the public schools were 'archetypal British institutions' (Boyd and McWilliam, 2007: 79).

At the core of this middle-class hegemony was the gentlemanly ideal, conveyed through cricket, and used – quite comfortably – for the advance of Britain's capitalist-imperial interests. The term 'gentlemanly capitalism' has been used to capture the dominance of city-based finance by the Victorian middle class (see Cain and Hopkins, 2007). This class was the 'class of the population that contributed least to [to the evolution of cricket]' yet it was 'destined to appropriate the game and convert it into a national institution' (James, 2000: 161). According to Mary Poovey, between the 1830s and 1860s a sizeable shift towards a single 'mass culture' was made (1995: 2). For Martin Weiner:

> By Victoria's death her nation possessed a remarkable homogenous and cohesive elite, sharing to a high degree a common education and a common outlook and set of values. This shared outlook represented an adaptation by the traditional ruling class (as was widely noted at the time), but it also marked a crucial rebuff for the social revolution begun by industrialization.
>
> (1999: 7)

This perpetual postponement of revolution was and remains crucial for Britain, enabling the hegemonising institutional middle class to maintain its own position, a position that is one of land-owning emulation. As Weiner states, the new ideal of gentlemanliness 'was in its essentials the older aristocratic idea purged of its grosser elements' and with it the 'entrepreneurial class', at their moment of hegemonic ascension, 'turned its energies to reshaping itself in the image of the class it was supplanting' (1999: 71). In many ways, this is why cricket has maintained its position as an institutionally-bound signifier of British paralysis, of a class-bound stasis enabled by the fallacies of the gentlemanly myth, the idealised version of imperial Englishness, and the mutations in relations that enable revolution to be re-encoded as stasis – or, as Tom Brown says, 'I'm all for Revolution and hurrah for Law and Order!' (Hughes, 1989: 171).

The grace of popularity

In the late Victorian sporting extravaganza, cricket enjoyed enormous popularity as well as its hegemonising strength of gentlemanly purpose (Bateman, 2009: 51). Across England, the game was solidifying into a county-based competition quite unlike its sporting peers. Although teams had been playing under county names – as with Kent – since the early eighteenth century, the structured creation of county cricket clubs was a nineteenth-century phenomenon with the formation of teams like Cambridgeshire, Lancashire, Middlesex and Yorkshire before Gloucestershire emerged with its Graces in 1870. As is often said of the late nineteenth century, the railways played a key role in enabling the movement of teams and thereby the possibility of regular games, including games between geographically distant counties. The newspapers also played a crucial role in publishing match results and scorecards, as well as feature stories about key players or performances, and by the 1870s they were also providing an informal table of county results for the season. The formal County Cricket Championship developed over a handful of years, between 1873, when the first player qualification rules where introduced, and 1879, when a formal arrangement was agreed to settle the way 'champions' would be decided – with the first official season beginning in 1890 and ending in the crowning of Surrey CCC. The tussle between North and South crystallised during the opening decade of the championship as Surrey and Yorkshire dominated the league title. It was also evident in the North versus South fixture played from the 1830s. All of the sixteen clubs in the championship by the turn of the twentieth century were English, with Glamorgan only entering the competition in 1921. Beyond participation, popular engagement with the sport depended on match day attendance and reading about the game in newspapers, popular sporting books and fiction, especially schoolboy and adventure fiction. Although county cricket has never been a financially sound system, gate money was significant and rising at the end of the century (Mangan, 1988a: 12), and sizeable crowds were evident at the popular annual fixtures and festivals, with Scarborough Festival running from 1876. Newspaper circulation brought cricket to many more than live games ever could, binding their readers together in a popular imagining of the cricketers and matches going on up and down the country and eventually overseas too. As Bateman writes, 'many more thousand read about W. G. Grace than watched him play' (2009: 16). Specific cricketing publications also emerged, with none more important than *Wisden*, created in 1864 and soon to become the 'bible' of cricket. Although initially *Wisden* reprinted old score cards and bound itself to the nostalgia of the always-already lost era of cricketing ruralism (Green, 1988: 57–58), it soon began publishing all results and reviewing each season, including its key moments and players, bringing to its softly conservative readership a sense of cricketing authority – domestic and international – that mirrored that of the MCC. On the back of cricket's growing popularity, John Wisden, himself an experienced first-class cricketer, had started selling cricket equipment in the 1850s in the Midlands before establishing the 'Cricket and Cigar' shop in London with Fred Lillywhite which mutated into the sporting equipment giant Lillywhites (ibid.).

This period of cricketing solidification and financially enterprising expansion was dominated by one figure – cricket's most iconic figure – W.G. Grace. Cricketing heroes had existed before Grace, but none had reigned over the game as Grace did. He was preceded in the mid-nineteenth-century popular imaginary by the controversies of Lord Frederick Beauclerk (1773–1850), a vicar-gentleman known for wagers, disputes and potential match-fixing, and the technical advancements of Fuller Pilch (1803–1870), a Kent player famed for his forward play (Green, 1988: 45), as well as numerous other stars of the game. Yet Grace became the cricketing presence of the century, responsible in no small part for massifying the game, disrupting its gentlemanly façade, pushing forward cricket's technical aspects, especially in batting, and helping establish the fearsome competition between England and

Australia that became the Ashes. He also amassed a multitude of statistical records, the most well-known being his status as the first player to accumulate 1,000 runs and 100 wickets in a single season. The basic facts of Grace's family life and cricketing success are often told, though his friendship with Charles Darwin is rarely reported (79). Suffice to say, the bearded bulk of a man who became a country doctor and handsomely rewarded 'amateur' cricketer came from a large family of Graces, was taught cricket by his mother, played the first-class game alongside some of his brothers – the most important being E. M., nicknamed 'The Coroner' to W. G's 'Doctor' – and used his county, Gloucestershire, as if it were a private entrepreneurial project (Green, 1988: 68). Stories of Grace's gamesmanship and sense of self-importance, as well as his ability to amass expenses, abound. Nevertheless, his popularity is incontrovertible. It has been said to account for the popularity of the Gentleman versus Players match, and to have raised sufficient gate money for Grace to be responsible for the building of 'half the pavilions in England' (ibid.). Few depictions of Grace are as enlivening and full of historical punch as those provided by C. L. R. James in *Beyond a Boundary* (1963). James' sense of the Doctor as a world-historical figure captures the importance of the cricketing colossus for Victorian England. To James, Grace was 'typical of an England that was being superseded', but rather than being a 'relic', or 'nostalgic curiosity', he was 'a pre-Victorian militant', bestriding 'two ages', and ensuring that a referent for 'preindustrial England [was found] in the iron and steel of the Victorian Age' (2000: 182). While we might see Victorianism as bound to the Queen's reign, we may be better off characterising cricket's popular Victorian transformation as occurring during the lifetime of Grace, from his birth in 1848 to his death in 1915, the year that the county championship was suspended because of the deathly costs of the Great War.

By Grace's prime Britain's empire was at its zenith, and cricketing culture had a distinctly imperial dimension. In varying capacities, official and unofficial, British people and the populations they came into contact with – of varying ranks and classes – influenced the diffusion of cricket and other imperial sports. Traders, military men, emigrants, educators, missionaries, administrators and local communities all played a part in establishing cricket in empire, as part of an ongoing process of imperial imposition, mimicry and opposition, as explained by Keith Sandiford and Brian Stoddart in *The Imperial Game* (1998). As the practice of Englishmen and other Britons abroad, cricket operated as a way of re-enacting their claim to Englishness, and its associated racial supremacy, in the face of the imperial 'Other'. Almost simultaneously, cricket became the target of imitation or emulation and an arena for politicised performances of competence, often as resistance. This resulted in the adaptation of imperial Englishness and its disciplinary discourses into localised racial, class and, in India, caste prejudices. As is often said, cricket became the sport of empire and the vehicle for anti-colonial self-expression, ultimately helping push claims for self-determination and independence. Participation in cricket for the colonies became the means of imperial allegiance and aggressive defiance. For 'white' colonies like Australia, New Zealand and South Africa, there was a relatively swift entrance into a cricketing and commonwealth collectivity of nations, meaning fierce sporting competition was played alongside a political union with, but under, Britain, represented by the England's cricket team. For 'non-white' colonial relations like India and the West Indies, however, the struggle for international cricketing acceptance and 'test' status was long and tainted by racism. Cricket was also being played in America and Canada and so early international tours in the 1860s were to North America. When the first visitors from Australia arrived in 1868, they were an all-Aboriginal team. England encountered the first all-white Australia team in the 1876–7 tour of New Zealand and Australia, organised by Lillywhite, and hosted a return tour in 1878. This constituted the beginnings of international 'test' cricket, resulting in the famous Australia series win in England in 1882 that inaugurated the Ashes, said to be cricket's oldest international rivalry.

* * *

The 'Golden Age' of cricket, roughly 1890 to 1914, moves us from the creation of county and international competition through to regular overseas tours and the dominance of the gentleman amateur as an administrative and political force. To catch a glimpse into this world of cricketing glory, four figures in particular can be said to highlight the importance of the pattern of institutional access and cricketing privilege that the mid-century public schools established. Born in Trinidad, Lord Harris (1851–1932) was an Etonian and Oxford blue, captain of Kent and England, Governor of Bombay (1890–95) and then-president of the MCC in 1895. Born in Gainsborough, Lord Hawke (1860–1938) was an Etonian and Cambridge blue, who captained Yorkshire and played for the MCC and England between 1896 and 1899, undertaking tours to India, America, South Africa, the West Indies, and other places. He was also president of Yorkshire for forty years and president of MCC during World War I. Ranjitsinji (1972–1933), the Indian Prince and Cambridge blue, played county cricket and for England in test matches between 1896 and 1902. His *Jubilee Book of Cricket*, published in 1897 for the Queen's 60th Jubilee, became a popular cricketing manual, offering players, especially schoolboys, technical advice with illustrations to help them improve. C. B Fry (1872–1956) was a Repton scholarship boy, then an Oxford blue, before playing for Surrey and England alongside his close friend Prince Ranji. He became famous for playing football – including in two FA cup finals – and competing in athletics alongside cricket, in which he scored abundantly. He taught at Charterhouse, accompanied Ranji to the League of Nations and stood as a Liberal candidate for Brighton in 1922. This was the generation of men leading cricketing through its 'Golden Age'.

These were the types of men representing cricketing heroics to the English public, and their attachment to British institutions is obvious and was determining. These institutional shadows hang over cricket and ensure that its Englishness is bound to the British state, a state strongly attached to nostalgic modes of self-narration.

References

Bateman, Anthony. (2009) *Cricket, Literature and Culture: Symbolising the Nation, Destabilising Empire*. Farnham, Surrey: Ashgate.

Bateman, Anthony. (2011) 'Cricket Pastoralism and Englishness'. In Bateman, Anthony., and Hill, Jeff. (eds.) *The Cambridge Companion to Cricket*. Cambridge: Cambridge University Press.

Baucom, Ian. (1999) *Out of Place: Englishness, Empire and Location of Identity*. Princeton, NJ: Princeton University Press.

Birley, Derek. (1999) *A Social History of English Cricket*. London: Aurum Press.

Birley, Derek. (2000 [1979]) *The Willow Wand: Some Cricket Myths Explored*. London: Aurum Press.

Bowen, Rowland. (1970) *Cricket: A History of its Growth and Development Throughout the World*. London: Eyre & Spottiswoode.

Boyd, Kelly and McWilliam, Rohan. (2007) 'Introduction: Rethinking the Victorians'. In Boyd, Kelly and McWilliam, Rohan (eds.) *The Victorian Studies Reader*. Abingdon, Oxon: Routledge.

Brookes, Christopher. (1978) *English Cricket: The Game and its Players through the Ages*. London: Weidenfeld & Nicolson.

Cain, P.J. and Hopkins, A.G. (2007) 'Gentlemanly Capitalism'. In Boyd, Kelly and McWilliam, Rohan (eds.) *The Victorian Studies Reader*. Abgindon, Oxon: Routledge.

Cardus, Neville. (1931) *Cricket . London: Longmans*

Chandler, Timothy J. L. (1991) 'Games at Oxbridge and the Public Schools, 1830–80: The Diffusion of Innovation'. *The International Journal of the History of Sport*, 8 (2). pp. 171–204.

Golby, J.M. and Purdue, A.W. (1999) *The Civilisation of the Crowd: Popular Culture in England 1750–1900*. Stroud: Sutton Publishing.

Green, Benny. (1988) *A History of Cricket*. London: Barrie & Jenkins.

Howarth, Patrick. (1973) *Play up and play the game: the heroes of popular fiction*. London: Methuen.

Hughes, Thomas. (1989 [1857]) *Tom Brown's School Days*. Oxford World's Classics Edition. Andrew Sanders Ed. Oxford: Oxford University Press.

Hughes, Thomas. (1861) *Tom Brown at Oxford*. London: Thomas Nelson.

James, C.L.R. (2000 [1963]) *Beyond a Boundary*. 5 Star Edition. London: Serpent's Tail.

Mangan, J.A. (1981) *Athleticism in the Victorian and Edwardian Public School*. Sussex: Falmer.

Mangan, J.A. (1987) 'Social Darwinism and upper-class education in late Victorian and Edwardian England'. In Mangan, J. A. & Walvin, James. (eds.). *Manliness and Morality: Middle-Class Masculinity in Britain and America, 1800–1940*. Manchester: Manchester University Press.

Mangan, J.A. (1988a) *Pleasure, Profit, Proselytism: British Culture and Sport at Home and Abroad 1700–1914*. London: Frank Cass.

Mangan, J.A. (1988b) *The Games Ethic and Imperialism: Aspects of the Diffusion of an Ideal*. London: Frank Cass.

Mangan, J.A. (1996) '"Muscular, militaristic and manly": The British Middle-Class Hero as Moral Messenger'. *The International Journal of the History of Sport*. 13 (1). pp. 28–47.

Marqusee, Mike. (1994) *Anyone But England*. London: Aurum Press.

Newsome, David. (1961) *Godliness and Good Learning: four studies on a Victorian ideal*. London: John Murray.

Nyren, John. (1974) *The Young Cricketer's Tutor*. Introduction by John Arlott. London: Davis-Poynter.

Poovey, Mary. (1995) *Making a Social Body: British Cultural Formation, 1830–1864*. Chicago, IL: University of Chicago Press.

Pycroft, James. (1854 [1851]) *The Cricket Field. 2nd Edition*. London: Longmans.

Quigly, Isabel. (1984) *The Heirs of Tom Brown: the English school story*. Oxford: Oxford University Press.

Richards, Jeffrey. (1988) *Happiest Days: the public school in English fiction*. Manchester: Manchester University Press.

Sandiford, Keith A.P. and Stoddart, Brian. (1998) *The Imperial Game: cricket, culture and society*. Manchester: Manchester University Press.

Vamplew, Wray. (1988) 'Sport and Industrialization: An Economic Interpretation of the Changes in Popular Sport in Nineteenth-Cenutry England. In Mangan, J.A. (ed.). *Pleasure, Profit, Proselytism: British Culture and Sport at Home and Abroad 1700–1914*. London: Frank Cass.

Weiner, Martin. (1999) 'Can Culture Explain Economic Decline?' In Boyd, Kelly., and McWilliam, Rohan. (eds.) *The Victorian Studies Reader*. Abingdon: Routledge.

Williams, Jack. (1999) *Cricket and England: A Cultural and Social History of the Inter-war Years*. London: Frank Cass.

Williams, Raymond. (1961) *The Long Revolution*. London: Penguin.

3 'Drivel for dregs'

Perceptions of class, race and gender in British music hall, 1850–1914

David Huxley and David James

Introduction

In 1906, W.T. Stead, a pioneer in the burgeoning field of investigative journalism, was persuaded to attend a performance at the London Pavilion Music Hall. Stead was deeply religious, and his most famous campaign, as editor of the *Pall Mall Gazette*, was an exposé on child prostitution in 1885 that had made him a nationally famous figure. His reaction to the music hall performance is revealing:

> My first impression was one of unutterable boredom. For three and a half solid hours I sat patiently listening to the most insufferable banality and imbecility that ever fell on human ears.
>
> (quoted in Anon., 1906: 3)

He continues to describe both the acts and their reception:

> Poor mimes, who kick up their heels on the stage and snigger and laugh at dirty jokes, the more degraded creatures in the stalls who applaud and murmur very good, very clever, as one piece of vulgar inanity succeeds another.
>
> (ibid.)

Thus Stead's attack is on both the performers and their audience. He finally sums up the performance as 'drivel for dregs'. The one act that Stead excused from his vitriolic attack was 'La Milo', the stage name of the star, Pansy Montague, who posed, painted white, as classical 'nude' statues. It might have been assumed that this 'pose plastique' would meet with his disapproval, but he described it as the one redeeming feature of the night. The nature of La Milo's act and the details of her successful career are described in Huxley (2013).

How did the music hall become the site of this vitriol, with its clear overtones of class tension? The ways in which the music hall became so controversial is indicative of the ways in which issues of class permeated almost all aspects of Victorian and Edwardian life in Britain. In this chapter, we will attempt to chart the way in which these changes developed, and using a little-known popular magazine, *Red Letter*, we will look at contemporary views of the music hall in the early twentieth century, and the surprising degree of freedom afforded to both female and black performers.

Class and the development of the music hall

From the latter part of the nineteenth century until the First World War, the music hall and variety theatre provided the premier form of entertainment in Britain. In 1900, there were at least 167 major music halls in Britain, but the terms used here indicate something of the complexity of this form of entertainment. 'The halls' underwent a slow development through 'penny gaffs' and 'song and supper clubs', through music hall and variety, and then onto a slow decline in the face of the gramophone record and, most significantly, cinema. The music hall began as a raucous, often rude, working-class entertainment but became a huge industry that created stars that could earn thousands of pounds in the currency of the day – millions in modern terms. The mixing of folk culture and the folk song with an increasingly organised capitalist structure of theatre venues is a prime example of 'compromise equilibrium' (Gramsci, 1971: 161), whereby a continual tension exists between arguably 'authentic' performers, and controlling owners and managers. The burgeoning music hall also provides a site that is not purely bound by issues of class, but also a fascinating resource in terms of gender and 'race', as it is an area where both female and black performers, despite prejudice, could escape traditionally proscribed roles. However, it is issues of class that tend to dominate discussions of music hall and its audience, and class divisions and definitions are a site of much debate. Class in Britain can be seen as a hierarchical continuum from the aristocracy to the street dweller at the bottom, or as a tripartite model of upper, middle and working class. Equally, class can be seen as a dualistic them/us patrician/plebeian model or, drawing on Karl Marx (1976), it can be seen as being defined by the relationship to the means of production. Although these models are discrete and arguably mutually exclusive, David Cannadine argues that, '...all of them are ignorant over-simplifications of the complexity of society. Yet they have remained remarkably enduring...' (Cannadine, 1998: 21).

As with many aspects of the music hall, the nature of the class of the audience is more complex than it first appears. In order to try and provide a more scientific approach, Dagmar Hoher has looked at the specific nature of audiences as revealed in a number of music hall disasters – fires, collapses and panics – the records of which reveal hitherto unknown details of the attendees. Thus this approach reveals that in a number of panics:

> In Dundee, 11 out of 13 and in Manchester two-fifths of those whose occupation was given were textile-factory workers: victims and witnesses

in Liverpool and Aberdeen a third of those whose occupation was given were seamen, ropeworkers and shore-labourers.

(1986: 80)

Hoher establishes that a range of occupations, ages and gender was represented, with more female attendees appearing in halls where the local industry – such as the textile industry of Manchester – employed more women workers. Equally, there were a huge range of different levels of workers from well-paid artisans to the poorest labourers, and different entrance fees – from 6d to 1d for different halls and the different parts of the hall – with private boxes being the most expensive section.

In order to try and 'define' – in some measure at least – the largely working-class nature of the audience, the work of Pierre Bourdieu is a useful guide, albeit in a somewhat simplified form. One of the ways Bourdieu characterises the various classes is through their tastes. He characterises the tastes of the working class as being formed out of necessity rather than choice, arguing that a virtue is made of this necessity. However, this necessity transcends its original practical purpose: 'The working class indicat[e] a preference for [the] realistic and practical… working men are forbidden every pretension' (Harker et al., 1990: 121). Class for Bourdieu is also oppositional; it consists (in some part at least) of what you are not. Bourdieu asserts that the tastes of the ruling classes 'can only assert themselves as such in relation to the tastes of necessity, which are thereby brought to the level of aesthetic and so defined as vulgar' (Bourdieu, 1986: 56). Vulgar here can be thought of in the senses of both common or demotic, and crude and coarse. It is exactly these senses of vulgarity that the music hall audiences seemed to love, and the artists knew their audience. It is also exactly what the music hall proprietors (and those that sought to censor the artists) tried to eradicate in their search for respectability. There was also great resistance to the mixing of the classes at entertainment venues, which was most clearly seen at popular seaside resorts as cheap train travel allowed the working classes to make excursions to the coast. John Walton argues that:

> During the nineteenth century the hegemony of the middle class values in the resorts was threatened on two fronts: by working class excursionists, not all of whom were thrifty and deferential and, later in the century, by the spread of an increasingly open, secular attitude to leisure activities, an apostasy from the strict controls of behaviour imposed by 'respectable' society at the mid-century.
>
> (1989: 134)

The solution for the resorts was to try and separate the classes geographically or by timing, but it was this very separation that the music hall proprietors were intent on dissolving. Walton points out that some resorts were able to limit 'excursionists' to Sunday rail trips (or even to ban such trips), whilst by 1862 Blackpool North Pier had a 'select lounge' that required a toll for entry, and Central Pier deliberately tried to attract the working class. The breadth of entertainment available to the Victorian public was huge, and encompassed everything from opera and serious plays, to melodrama, circuses, music hall, variety theatres and street performers. At the same time, the music hall did not operate in isolation, and could influence the 'legitimate' theatre. As Hugh Cunningham points out: 'Not only were individual performers rising to fame, but whole types of performance, popular in origin, were having a marked impact on high culture' (1982: 67)

Cunningham cites the example of the increasing use of animals in theatre productions at Drury Lane and Covent Garden by the 1850s, but perhaps the apotheosis of this upward mobility was Edward Moss's establishment of The Royal Variety performance in 1912. Moss, knighted in 1905, was a key figure in the music hall's search for respectability. Born in Manchester in 1852, the son of a music hall performer, he formed Moss Empires Limited in 1899 with Oswald Stoll which became the largest and most important of the theatre chains in the UK. The chain enforced strict censorship, and for the first Royal Variety Performance (then called The Royal Command Performance) they even excluded the biggest star of the day, Marie Lloyd, because her material was deemed to be too risqué for King George V and Queen Mary.

From folk song to sheet music

Music hall songs – the mainstay of the halls' appeal, along with comedy – stemmed from a folk tradition. The problems of retrieving the 'true' folk song are described in Harker (1985), which demonstrates that the concept of 'authenticity' is highly fluid. But inevitably there were significant changes as these folk songs moved from the country to an urban setting. The bawdy folk song has a long and almost noble tradition and has been utilised (or sometimes written) by authors as diverse as Chaucer, Shakespeare and Burns. Hundreds of these songs, normally by anonymous authors, moved gradually from the amateur sphere to more organised venues. The way in which this happened, with street performers and country fairs gradually becoming less important, is probably impossible to detail at this distance. Waites et al. argue that:

> Popular culture entails a system of production, although that is not all it entails and pinpointing its historical origins in Britain is not easy. Until recently, many historians would have accepted the idea of a 'Bleak Age' for the common people separating the popular pursuits of eighteenth century largely rural, society from the new pattern of working-class leisure of urban mid-Victorian Britain. Such an idea fitted the pessimistic interpretation of the social consequences of early industrialisation. But what is certain

is that the early unregulated forms of entertainment informed the later commercial forms of entertainment in quite a direct way.

(1986, p.15)

Yet they continue to describe:

> A thriving popular culture dating from the 1820s, based on fairs, shows and travelling theatrical troupes, supplied by thrusting cultural entrepreneurs and common to rural and urban society.
>
> (ibid.)

The change from a 'folk art' into an organised controlled industry was slow but, in the end, dramatic. An important step was the charging of admission – even if it was only a penny, which gave the 'penny gaffs' their name. H. Chance Newton, a veteran musical critic for *The Referee* under the pseudonym 'Curados', emphasised the importance of the pub origins of the music hall:

> In those days every music hall was but the chief room or at best, a mere annexe, of this or that tavern or gin palace. They were run entirely on drinks, or what is usually known as 'Wet Money'. Indeed, at some admission was by at so-called 'refreshment ticket', half of the cost of which was given back to the patron in the form of liquids.
>
> (1928: 1)

Liquor sales continued to be important, if not vital, to the financial success of the halls. This, however, fuelled the opposition to them from the strong temperance movement that flourished in nineteenth-century Britain. Although the methods of temperance crusaders varied, many were inspired by religious beliefs such as the missionary Frederick Charrington, who was ironically the son of a famous brewing family. In the 1880s, Charrington picketed several London music halls with such zeal that he was arrested. In 1885, 'Charrington was again in court charged with slander and libel of a music hall owner and an injunction was issued against his harassing techniques. He was found guilty, having called Lusby's music hall "a pit of hell"' (Shiman, 1988: 126).

The nature of the performances in the 'song and supper rooms' is demonstrated in Thackeray's 1855 novel *The Newcomes*. In the novel, Colonel Newcome goes to a real venue that the author himself had visited, the Cave of Harmony in Covent Garden. The upstanding Colonel is surprised at the nature of the entertainment:

> Here Colonel Newcome brought his young son in the expectation of enjoying an evening of glee singing and old English ballads, but when later in the evening Captain Costigan rose to his feet and gave voice

> to a song, it so offended the worthy Colonel that he called for silence. When some in the room urged the singer to go on, the Colonel retorted, 'Does any man who has a wife and sisters, or children at home, say,' Go on' to such disgusting ribaldry? Do you dare sir, to call yourself a gentleman, and to say that you hold the King's commission, and to sit down amongst Christians and men of honour, and to defile the ears of young boys with this wicked balderdash?
>
> (quoted in Speaight, 1975: 7)

A small number of these songs were printed from the 1830s onwards, another important step in the commercialisation of the form, and some of the authors and singers are known by name. Although heavy with euphemisms, these were performed in comparatively control-free spaces and were much less subtle than the later music hall songs. These pamphlets were produced by publishers of pornographic material, and there is no attempt to hide the nature of the songs; for example, the lyrics of 'The Wonderful Belly Physic' is introduced with the boast 'A Capital New Smutty Song'. The barely disguised nature of the lyrics is demonstrated in 'Sally's Thatched Cottage' where the ninth verse runs:

> Said Sal, my door was tight and close,
> It would scarcely admit a mouse,
> But now it opened so often has been,
> A donkey might with ease get in.
>
> (quoted in Speaight, 1975)

Some songs also have a more blatant style that would not survive into the era of the music hall, such as 'The Way to Come over a Maid', which ends:

> If you seduce a maid,
> You must swear, and sigh, and flatter,
> But if you'd win a widow,
> You must down with your breeches and at her!
>
> (ibid.)

These kinds of songs provide a link back to bawdy folk traditions that could not survive the restrictions of a more inclusive formalised entertainment.

For the successful music hall act with a popular song there was not just the potentially rich rewards of their wages in halls across the country, there was also the possibility of royalties from the sale of sheet music. Victorian homes of many different classes had musical instruments, often pianos, on which they could provide their own entertainment. Whilst some homes might favour hymns, others would want to recreate the latest 'hit' song. In America, publishers had been successful through most of the nineteenth century and by 1893 could reputedly sell over a million copies of an individual song like 'After the Ball'. In Britain, Francis, Day and Hunter (formed in 1874) were the most successful of these publishers, and specialised in music hall songs. Royalties paid to composers and artists from a successful song could thus provide a sizeable adjunct to incomes from performing in the music hall.

In 1889, the recording system of Edison's phonograph cylinders provided homes with an actual recording of their favourite artists, and this would change the nature of the unique performance in the halls in a significant and lasting way.

Industry and regulation

The containment of these performances in various venues such as the the Cider Cellar and The Coal Hole in London created liminal spaces that were, at this stage, largely a male preserve. Now that entertainment venues had been created, there was an increasing elaboration of performances, the establishment of a stage and more formal seating arrangements. In the search for greater profits, the proprietors of these venues tried to widen the appeal of their performances without destroying the broad appeal they had already established. This led to a long and sometimes bitter battle between, on the one hand, owners and censors, who wanted increasingly safe and respectable entertainment and, on the other hand, performers and audiences, who revelled in risqué material. The position of proprietors was particularly delicate, as they knew perfectly well that much of the appeal of their acts was their risqué nature, yet it was also this material that might offend the wider audience they hoped to attract.

One key to this was making these spaces acceptable to a more middle-class audience, and, in particular, encouraging a female clientele. Music hall pioneer Charles Morton, owner of the Canterbury Hall, introduced 'Ladies' Thursdays' in the early 1850s, ostensibly to mitigate against the raucous behaviour of the male audience, but this also encouraged the association of the halls with prostitution. In 1894, campaigner and author Mrs Ormiston Chant objected to the performances at Empire Theatre of Varieties in London and, more importantly, the prostitutes who frequented the promenade section of the theatre. The theatre was briefly closed, but opponents of Ormiston Chant (which included a young Winston Churchill) succeeded in having it reopened. Ormiston Chant has often been presented as an outmoded 'blue stocking' critic of the halls, but her campaigning against the evils of prostitution, along with other campaigners such as Stead, was much more serious. The number of prostitutes in London in 1870 is disputed, but it has been estimated at anything between 6,000 and 80,000, the latter figure probably an exaggerated estimate by the Bishop of Exeter, quoted in the contemporary research of William Acton. Certainly prostitution was very widespread, and it was in fact not just prevalent in music halls, but also in the 'legitimate' theatre. Puckler-Müskau, a German prince, visited London in the 1820s and commented:

> A second reason for the absence of decent families from the theatre is the attendance of several thousand filles de joie...during the intermissions they crowd the large and fairly elaborate foyer, where they put all their effrontery unrestrainedly on show. It is strange that such spectacles are in no country on earth more shamelessly displayed than in pious and decent England.
>
> (quoted in Fisher, 2001: xxiii)

Nevertheless, the spectre of prostitution haunted the reputation of the halls, and even influenced the status of female performers, who were often regarded as less than 'respectable'. Sara Maitland describes the problems of public performance for women:

> By the second half of the century 'decent' women were supposed neither to have nor to desire a public life...the actress who not only displayed herself in public, but did so for money, was the very antithesis of the Victorian 'Lady'.
>
> (1986, p.65)

But, the female performer in patriarchal society nevertheless wielded a power unknown to the majority of women. Those who disapproved of the female performer did not understand, or resented, this power. As Faulk points out:

> The social purity movement, adept at marshalling the media to defend women against rapacious male sexuality, had difficulty in imagining how women performers could be the agent rather than the subjects of their publicity.
>
> (2004, p.186)

The role of female performers

A small number of the smoking room songs are clearly meant to be sung by a female performer – which suggests that the early 'male space' might not have been quite as exclusive as it is normally supposed to be. The status of any female performer in such an environment in the early nineteenth century may well have been questionable, but this also indicates the problem of retrieving the detail of popular entertainment history at this distance. Even well-documented female performers are subject to myth and rumour. There is a story that Marie Lloyd, under censorship pressure, changed the lyrics of a song from 'She sits among the cabbages and peas' to 'She sits among the cabbages and leeks'. This appears in many sources, including her Wikipedia entry, despite the fact that Farson proved that the story was apocryphal as early as 1972. Lloyd, however, was a major star – arguably the biggest star – that British music hall produced. Her status, and the existence of hundreds of other female performers, many of them hugely successful, calls into question the whole issue of the position of women during this period. Whilst there may have been huge restrictions on the roles of women in mainstream society, there is also clear proof that for the female performer, the music hall provided a freedom, both in the way she could act and also in her financial power that may have been unavailable outside this sphere. Also, Lloyd was by no means unique – extant posters and records at the National Fairground Archive at Sheffield University prove that there were hundreds of female comics and singers, many of whom, such as Bessie Bellwood and Vesta Tilley, were major, well-established stars. Tilley was the foremost amongst male impersonators, and a major star between 1870 and the First World War. Although very careful to stress her traditional 'femininity' in some publicity photographs, her performance on stage was another example of the freedom for women to ridicule traditional masculinity. This could not be taken too far, as Maitland points out:

> For male-impersonators to survive, they have to find a supportive community of men; they cannot hold those men up to the same sort of ridicule that it is possible for men to hold women up to because they are dependent on them.
>
> (1986: 101)

Nevertheless, her act often consisted of satirising foolish and deluded male 'swells'. The records of the Argyle Theatre in Birkenhead also reveal the female performers could earn just as much as male stars.

Red Letter was a popular magazine published by D. C. Thompson. Positioning itself as a mainstream publication with wide appeal, it featured romantic or crime fiction, sport reports and entertainment reviews. In the early years of the twentieth century, the magazine ran a 'how I made it to stardom' strand for various music hall stars. One of these from February 1914 is titled 'Florrie Forde lets you into a secret and tells how she climbed to the top of the bill'. Forde, despite being born in Australia, was by 1914 one of the great stars of the British music hall. She popularised songs such as 'Down by the Old Bull and Bush' and, during the First World War, 'Pack up Your Troubles' and 'It's a Long Way to Tipperary'. She begins the article by saying how she gets many letters each week from women asking how to 'become a celebrated footlight favourite' (*Red Letter*, February 17, 1912: 182). She divides these women into two categories; first, there are the:

> Illiterate young women, disgusted by the daily drudgery of their days, and attracted by the glamour of the bright lights, the beautiful dresses and the merry music of the stage, to long for a life of luxurious ease and all-conquering successes. Such are the dreams of ignorance that are fated to become nightmares when the dreamer awakes to realities.
>
> (ibid.)

There was also the other category though:

> The letter before me, however, is not of this type. Both the handwriting and the phraseology prove it to be the letter of a well-educated girl whose head is screwed on the right way and who recognises that to make her way to the top of the profession, she must build a solid foundation and begin at the bottom of the ladder. She desires to join me in order to learn the rudiments of the business, and to become first of all a dancer.
> (ibid.)

Clearly, these two categories are divided along class lines. The uneducated and illiterate young woman is working class, 'disgusted by the daily drudgery' of her life. The implication seems to be that the working-class women expect instant success, whereas the more educated (middle-class) woman whose 'head is screwed on the right way' recognises that she needs to work her way up from the bottom. Interestingly, this corresponds to the trope of the working class being associated with instant gratification and the middle classes deferring gratification for a later but greater reward. She goes on to tell the reader that she herself has risen through the ranks from the chorus through to a ballad singer to musical comedy, a legitimate actress and a popular pantomime principal boy, and is thus qualified to give out advice on the subject. Her advice to the young lady on this occasion is that she cannot take her on, but that she should write to Mr Tiller and – if possible – place herself under his tutelage and join one of his dance troupes.

Forde herself was from a *petit bourgeois* background. Her mother was a theatrical costumier who had been a performer. What Forde is effectively saying is that in order to be successful, then hopeful young women should be like she was: presumably well educated and comfortably well off.

Similarly, Gertie Gitana (best remembered as the singer of 'Nellie Dean') stresses the hardship of a career on the stage, while noting that 'I had a good home with my uncle and aunt and could afford to wait until an engagement came up' (*Red Letter*, March 30, 1912: 350), and Florence Baines (largely forgotten now but a star in her day) stresses that although her first public engagement was singing in Blackburn Market Square, 'we were not so badly off at home that I had to help eke out the income in this way' (*Red Letter*, 1912: 98). Like Forde and Gitana, she goes on to stress the precarious financial nature of the profession, and describes her rise from 1 shilling a week to 2/6, to her sharing top billing with Dan Leno.

None of these female performers seem to have come from a particularly poor background, whereas some at least of the male performers make a particular point of the poverty of their background. For instance, George Formby Sr, in an article entitled 'My Life when Singing for Coppers: George Formby Tells How He Rose from Street Arab to Star Comedian' begins 'Mine is not an isolated story of the stage. Which of the 'bhoys' I wonder, can say they never knew days of privation and distress? But of them all, which of them, I wonder, can tell a more pathetic story than I?' (*Red Letter*, November 25, 1911: 322–3).

The role of black performers

The performance sphere in the late nineteenth and early twentieth centuries was increasingly international. Successful British acts could expect to tour not only the Empire (in particular South Africa and Australia) but also Europe and the USA. Importantly, American performers were also highly successful in Britain, and American influence permeated British acts. American 'nigger minstrel' shows had been popular in the USA since at least the 1840s, and Christy's Minstrels appeared in the UK in 1859. This group helped to establish the 'comedic' opportunities of white performers in 'blackface', and introduced minstrel songs and the 'cakewalk' dance. American influence was often seen as unfortunate by many British commentators, and some tried to create a backlash to trends such as the craze for ragtime music at the turn of the century. Yet 'blackface' performers were welcomed and imitated by the British. We have discussed elsewhere the inherent racism in much music hall performance (Huxley and James, 2012). Blackface comedians or singers, often, but not always white in reality, tended to pander to the worst stereotypes that were common in all popular culture of the period. Chirgwin, 'the White-Eyed Kaffir', G. H. Elliot, the 'Chocolate Coloured Coon', Eugene Stratton and many lesser stars performed acts that would seem offensive to modern audiences, and indeed by the early twentieth century there was already concern about these portrayals. Yet for a small number of highly paid black performers, often American but performing in the British halls, there was a similar sense of freedom and esteem that was also available to female performers. This is not to say that in the performance sphere they were able to escape pandering to the same racial prejudices.

However, probably the clearest example of the possibilities offered to black performers in the British music hall is shown in the careers of Eddie Whaley and his partner Harry Scott, who came from the American vaudeville circuit to appear in Britain in 1909. Whaley and Scott were featured in the weekly feature on music hall acts in the *Red Letter* of January 10th 1914. An indication of the endemic racism that permeated British society is encapsulated in the first few lines of the review 'Until I saw Whaley and Scott': the author states, 'I was never a keen enthusiast whenever coloured men attempted to create humour' (*Red Letter*, January 10, 1914: 39) The clear inference is that black performers are not funny, though Whaley and Scott are obviously exceptions here. The act was, in many ways, a standard double act with Whaley's 'Cuthbert' character acting as the straight man to Scott's 'Pussyfoot', the bumbling, foolish comic. This smart/foolish relationship is manifested in several ways. Whaley is dressed immaculately in evening wear whereas Scott's clothes 'suggest something that has just been imported from the tropics – the real negro' (ibid.). Whaley's speech, as transcribed in the magazine, is not obviously 'black' whereas Scott's clearly is. The article attempts to capture this with dialogue such as 'Wait till I ketch you alone' and 'Oh, don't Aa?' (ibid.). What the article fails to mention (perhaps failed to notice?) is that Scott performed in blackface, presumably to emphasis his simple stage persona. In fact, there is a fragment of film from 1933 that gives a flavour of their act where they engage in a bit of comic dialogue and end with a song. This short film does give an idea of their 'Pussyfoot and Cuthbert' stage personas. The idea of a black performer in blackface seems beyond absurd now, but as the entertainer Adelaide Hall observed:

> In the old days, black comedians like Bert Williams, who starred in the Ziegfield shows, were expected to black up. If they appeared on stage without it, they were booed off! Josephine Baker blacked up at the start of her career. Florence Mills and Paul Robeson were among the first performers to be accepted on stage without having to wear blackface. They broke away from tradition.
>
> (Hall quoted in Bourne: 2001: 2)

Hall is here referring to American performers and audiences, and it is perhaps no coincidence that all of the performers named by Hall (as well as Hall herself) successfully toured in Britain and Europe where racial segregation, as practised in America at the time, was all but unknown (though racism per se was clearly endemic). With one blacked-up performer and one elegantly dressed, Whaley and Scott perhaps managed to navigate any racist expectations on the part of the audience. Having first come over in 1909, they had continued to tour the halls and by 1914 – the time of this review – they had become UK residents. They toured successfully for many years and went on to appear regularly from 1933 onwards in the BBC radio series 'Kentucky Minstrels'. They starred in a film of the same name in 1934, becoming the first black performers to star in a British film. They went on to feature in a further five films; *On the Air* (Herbert Smith, 1934), *Minstrel Days* (John Baxter, 1934); *Pathetone Parade* (Fred Watts, 1936); *Shooting Stars* (Eric Humphriss, 1937) and *Take Off That Hat* (Eric Humphriss, 1938). Their partnership continued until 1946; Harry Scott died in 1947 and Eddie Whaley retired to run a hotel in Brighton shortly afterwards, where he died in 1960. Their highly successful career in the UK was arguably unattainable in America because of the de facto apartheid conditions, even in the north. However, they clearly also had to navigate the endemic racism in the UK, especially in Scott's 'Pussyfoot' character.

Conclusion

The music hall was accidentally complicit in its own decline as most music hall venues began to show films as part of the evening's entertainment from the 1890s onwards. 'Bioscope' operators, often near the bottom of the bill, showed short documentary films. These short films gradually moved up the bill, and by the early twentieth century some theatres converted into cinemas. The Victoria Theatre in London became the Victoria Picture Theatre in 1907. Inevitably, purpose-built cinemas followed, such as the Electric Theatre in York in 1911. The music hall and variety as live performance continued its decline, but its virtual demise by the 1950s led to an almost instant nostalgia, with the BBC screening the music hall tribute programme, 'The Good Old Days' from 1953 to 1983, filmed in the Leeds City Varieties Hall (built in 1865). The wider influence of music hall is perhaps less evident. Clearly it provided two of the famous stars of early cinema in Charlie Chaplin and Stan Laurel, and later comedians such as W. C. Fields and Will Hay, all of whom honed the skills learnt in the halls during their cinematic careers. By the 1950s, music hall, in the shape of variety, migrated to TV with hugely successful shows such as 'Sunday Night at the London Palladium' as well as less famous shows such as 'The Music Box'. It survives today in 'Britain's Got Talent', and the like.

Although the work of many of these music hall acts has disappeared, a more subtle influence remains in popular music, with performers as diverse as The Beatles, The Sex Pistols and Ian Dury of The Blockheads having expressed their debt to the British music hall tradition, recognising, perhaps, the irreverent nature of this form of popular culture. We would also argue that artists such as Lenny Henry and Jo Brand owe a debt of gratitude to the pioneers that made their living in the halls a century ago.

References

Acton, W., 1857 (1870) *Prostitution, Considered in Its Moral, Social, and Sanitary Aspects, in London and Other Large Cities and Garrison Towns, with Proposals for the Mitigation and Prevention of Its Attendant Evils*. London: John Churchill and Sons.

Anon., 1906. Drivel for Dregs. *The Queanbeyan Age*, 7 Dec, p.3.

Bourdieu, P., 1986. *Distinction: A Social Critique of the Judgement of Taste*. Trans. Nice, R. Cambridge, MA: Harvard University Press.

Bourne, B., 2001. *Black in the British Frame: The Black Experience in British Film and Television*. New York: Continuum International Publishing Group Ltd.

Cannadine, D., 1998. *Class in Britain*. New Haven, CT: Yale University Press.

Cunningham, H., 1982. Class and Leisure in Mid-Victorian England, in Waites, B., Bennett, T. and Martin, G. (eds) *Popular Culture: Past and Present*, pp. 66–91. *London*: Croom Helm.

Farson, D., 1972. *Marie Lloyd and Music Hall*. London: Tom Stacey.

Faulk, B. J., 2004. *Music Hall and Modernity: The Late-Victorian Discovery of Popular Culture*. Athens, OH: Ohio University Press.

Fisher, T., 2001. *Prostitution and the Victorians*. Stroud: Sutton.

Gramsci, A. 1971. *Selections from Prison Notebooks*. London: Lawrence and Wishart.

Harker, D., 1985. *Fakesong: The Manufacture of British Folk Song, 1700 to the Present Day*. Oxford: Oxford University Press.

Harker, R., Mahar, C. and Wilkes, C. (eds), 1990. *An Introduction to the Works of Pierre Bourdieu: The Practice of Theory*. Basingstoke: Palgrave Macmillan.

Hoher, D., 1986. The Composition of Music Hall Audiences, 1850–1900, in Bailey, P. (ed.) *Music Hall: The Business of Pleasure*, pp. 73–92. Oxford: Oxford University Press.

Huxley, D., 2013. Music Hall Art: La Milo, Nudity and the Pose Plastique, 1905–1915. *Early Popular Visual Culture*, 11(3): 218–236.

Huxley, D. and James, D., 2012. No Other Excuse: Race Class and Gender in British Music Hall Comedic Performance, 1914–1949. *Comedy Studies*, 3(1): 17–28.

Maitland, S., 1986. *Vesta Tilley*. London: Virago Press.

Marx, K., 1976. Preface and Introduction to *A Contribution to the Critique of Political Economy*. Peking: Foreign Languages Press.

Newton, H. C., 1928. *Idols of the Halls, Being my Music Hall Memories*. London: Heath Cranton Ltd.

Shiman, L. L., 1988. *Crusade Against Drink in Victorian England*. Basingstoke: Macmillan.

Speaight, G. (ed.), 1975. *Bawdy Songs of the Early Music Hall*. Newton Abbot: David and Charles.

Waites, B., Bennett, T. and Martin, G. (eds) (1986) *Popular Culture: Past and Present*. London: Croom Helm.

Walton, J., 1989. Residential Amenity, Respectable Morality and the Rise of the Entertainment Industry, in Waites, B., Bennett, T. and Martin, G. (eds) *Popular Culture: Past and Present*, pp. 133–145. *London*: Croom Helm.

4 Reading historical photographs

Class and gender in nineteenth-century images of Wigan pit brow women

Sarah Edge

In *Camera Lucida* (1984), Roland Barthes argued that the nineteenth century 'invented History and Photography' (1984: 93). In this remark, he suggests that both types of new 'information' should be approached as 'inventions'; that is, discourses created with an intention in mind. One, to shape the concept of the past, 'the fabrication of memory according to positive formula' (93) and to undertake the ideological work of naturalizing the present in which 'the stress on human *self development*, history ... loses its exclusive association with the past and becomes connected not only to the present but also to the future' (Williams, 1981: 119), while the 'age of the Photograph' was 'also the age of revolutions, contestations, assassinations [and] explosions' (Barthes, 1984: 94). This setting suggests a link between the invention of a new and unique way of picturing the world and the upheavals of the Industrial Revolution, and the new identities it spawned. This chapter will consider this relationship through a detailed reading of one of the very first uses of photography in a popular cultural context: a series of photographs of women miners who worked on the pithead of the mines in the Lancashire town of Wigan, made public for sale in the 1860s (Edge 1998).

Louis Daguerre in France and Henry Fox Talbot in England had made the invention of photography public via the press in 1839, though the first photographic image is usually cited as made by Joseph Nicéphore Niépce in 1826–7. Both announcements emerged from experimentation in the area of science and technology by prosperous men. The primary drive was to devise a process to permanently fix the image generated by light travelling through the artist's tracing tool, the camera obscura. It was their success in permanently 'fixing' an image created by light touching an object in the material world that allowed the iconic sign system of photography to become the most forceful ideological tool in the nineteenth century. It appeared to be a mechanical chemical tracing of the material world, but highlighted the 'historical paradox' of modernity which was 'the more technology develops the diffusion of information (and notably images) the more it provided the means of masking the constructed meaning under the appearance of the given meaning' (Barthes, 1982: 46).

Photographic theory, characterized by the work of Roland Barthes, drew upon linguistic theory to examine the distinctiveness of photographic communication (Edge, 2013). Identifying how the photograph is made up of two signifying systems – the indexical sign and iconic sign system – impacts on how the photographic message works (a digital photograph is still predominantly made via this process of electronically capturing light touching something in the real). In a photograph of a person, the indexical part of the sign is made via light touching the body and creating an individual and distinctive trace unique to all humans, similar to a fingerprint. The chemical printing process permanently fixes this trace, transforming it into an iconic sign system, a sign that looks like the thing it represents. Barthes identifies this as a powerful form of denotation, 'the scene *is there* captured mechanically, not humanly' (1982: 44, original emphasis). The photograph appears to be an image made, for the first time, free from human corruption, purer than its predecessor, the sketch, 'for there is no drawing without style' (43). Connotations, the cultural meanings of denotation, attach themselves to the iconic codes caught in the camera's action. These include everything in the frame: pose, fashion, props, backdrops, location, etc., as well as everything related to the act of photographing: lighting, camera angle and even what is selected to be photographed (20–25). It was this unique signifying structure 'invented' in the nineteenth century that allowed the photograph to undertake unprecedented ideological work because 'the mechanical' quickly became regarded 'as a guarantee of objectivity' (44).

Victor Burgin's overview of Barthes explains how, when analysing the (now infamous) photographic cover of *Paris Match* of a black solider giving the French salute (Barthes, 1973), Barthes discerned 'an additional message at work here' in which 'the very "naturalness" of the photograph ... lends the ideology ("myth") of imperialism the status of an indisputable truth' (Burgin, 1986: 75–6). In this approach, the ideological work undertaken by a photograph is made transparent. Under this methodology, exposing what a photograph meant or means is inextricably linked to studies of popular culture because it is the reader's popular cultural knowledge that activates the connotations caught in the photographic trace. Barthes explains how all photographs are 'historical' because they 'depend ... on the reader's "knowledge" just as though it were a matter of real language [*langue*], intelligible only if one has learnt the signs' (1982: 28).

In this context, reading a photograph of the past is particularly complex as the popular cultural language that activated its meaning has often been lost, a problem obscured by contemporary culture's tendency to employ historical photographs as a source of evidence of the past as illustrations of how things used to be. However, the methodology of critical photographic studies draws constructively from Barthes' semiotic approach to affirm that 'histories are not backdrops to set off the performance of images' because photographs 'are never "evidence" of history; they are themselves the historical' (Tagg, 1988: 65). In the nineteenth century, such an approach highlights the interrelationship between the 'invention' of both history and photography, confirming that both need to be approached as discursive forms implicated in the creation and naturalization of the unequal relationships of power that marked the period.

Nevertheless, from a popular cultural perspective, research on photography is restricted to the type of photographs that survived. For quite practical reasons, these tend to be located in two areas: first, official institutional uses of photography that survived because they were archived by powerful institutions. Studies reveal how these were linked to the rise of institutional discourses of power, identified in the work of Michel Foucault (1973, 1979, 1981 and 1986). An impressive body of work confirms this ideological role: see medical photography (McGrath, 1984, 2002; Tagg, 1988; Green-Lewis, 1996), prison photography (Tagg, 1988; Sekula, 1992; Green-Lewis, 1996), anthropological/ethnographic photography (Green, 1984; Edwards, 1992; Ryan, 1997) and charity or schools photography (Tagg, 1988). Nonetheless, from a wider popular cultural studies perspective, these photographs did not generally circulate within the terrain of popular culture and were instead viewed in a more exclusive institutional context.

The second source is early domestic photography that survived, archived in the private albums of the emerging industrial middle classes. The first commercial photographic daguerreotype studio opened in London in 1841 (Heathcote and Heathcote, 1979), and by the 1850s photography was available for the wealthy on the high street (Linkman, 1993). Contemporary scholarship confirms that photography's popular commercial success was linked to the desire of the emerging bourgeois to consolidate a sense of identity (see Hirsch, 1981; Spence and Holland, 1991; Linkman, 1993; Mavor, 1996; Smith, 1998; Holland, 2000; Tagg, 1988). This was initially achieved by drawing upon the iconic codes of self-portraiture operating in fine art. However, as the new medium took off, high-street photographers established their own professional practices to communicate class-based meanings. For example, professional manuals began to describe how sitters should be posed – 'a lady should not have that boldness of action which you would give to a man, but be modest and retiring, the arms describing the gentle curves and the feet never apart' – while men could be allowed 'greater assurance and assertiveness ... crossed legs, elbows at angles, with canes and umbrellas projecting into the space around them' (Linkman, 1993: 46). In this context, the class with the economic power to invest in their own self-representation has left modern scholars a trace on how they began to use the new medium for their own ends.

In this determining context, photographs of other groups of people are extremely rare, and my research is indebted to a quite extraordinary archive left behind by a middle-class urban man, Arthur J. Munby (1828–1910). Munby was a diary writer, note taker, collector, artist and poet who died in 1910, bequeathing two deed boxes now held at Trinity College, Cambridge to remain closed until 1950. When opened, they contained his sixty-four diary volumes; twelve notebooks; twenty-nine volumes of '*Visits to Hannah* 1885–1907'; Munby's two-volume *Life of Hannah*; twelve notebooks on Hannah (Hannah Cullwick, his working-class partner, a maid-of-all-work who he had secretly married in 1873); and other miscellanea. In this varied information, Munby left a record of his fascination with working-class women that has been examined by wider mid-Victorian scholarship (Hudson, 1972; Hiley, 1979; McClintock, 1995; Reay, 2002). However, somewhat overlooked by photographic scholarship (Edge, 2016) is that the deed boxes also contained hundreds of the earliest commercially produced photographs of working-class women collected by Munby: photographs of women performers from the early music halls; very early tourist photography from Europe; over seventy examples of the first uses of domestic photography by female working-class servants, maids and factory workers; internal tourist photographs from the Yorkshire Coast, chiefly of Whitby and Filey fisherwomen; and, finally, well over a hundred photographs of the Wigan pit brow women collected between 1860 and 1887.

The 'evidence' Munby offered on the first uses of photography in a popular working-class context, when added to the domestic uses by the bourgeois, identifies how quickly photography became a medium of popular mass communication. This allows scholars to locate the photograph as the first visual media that emphatically allowed for the mass distribution of images of the self and others, linking it to studies that examine how the modern mass media created the 'imaginary' identities of capitalism (Anderson, 1991). Therefore, the 'invention' of the popular press allowed the new industrial bourgeoisie to 'visualize in a general way the existence of thousands and thousands like themselves through print-language... Thus in world-historical terms bourgeoisies were the first classes to achieve solidarities on an essentially imagined basis' (1991: 74), confirming that neither the bourgeoisie nor the working class could have 'existed' without the invention of modern communication systems of which photography, with its 'peculiar' signifying system, was without doubt the most powerful.

Victorian pit brow women: photography discourse and power

Munby's fascination with pit brow women preceded his purchasing of the photographs in his archive. He had been visiting the Wigan pits from at least 1853 onwards, to sketch and interview the women. However, in 1860, he:

> went out at 9.30 to Mr. Craig the photographer, a very civil man, about obtaining a photograph of a pit woman in costume ... presently one of those well-known figures came up the street – a woman in flannel trousers, waistcoat (with lively buttons, in this case) and pink skirts, and a lilac cotton bonnet. A very comely woman about 30, married. She was clean because 'her dirty work had not begun yet'.
>
> (29/9/1860)

Craig, the photographer, informed Munby that he 'had photographed Elizabeth Fairhurst before'. Elizabeth agreed to the request for a photograph of her in working clothes but she 'was very anxious not to be exhibited in the window: "if my fella should see it, he'd about kill me"' (29/9/1860). In this account, Munby alludes to a number of details about this nascent industry. First, that Elizabeth Fairhurst had been photographed before; second, that photographs of pit women in their dirty working clothes were being exhibited for sale in the cheap photographer's windows; and, finally, that such a practice was interpreted as not respectable.

To begin the process of reconstructing what this account means and how it relates to what the photographs meant, I will start by examining what has been recorded in the camera's indexical action and transformed into iconic denotation: a woman; clothes; dirt; stances; props; backdrops; and location, inside or outside. To excavate the meaning attached to each part of this denotation requires a research process which moves outside the photographic frame itself to undertake a historical and cultural rebuilding of the photograph's connotations for the mid-Victorian viewer. This is a time-consuming process that Holland aptly describes as 'an act of decoding or historical detective work' (2000: 121) because, as Sekula's (1982) work on photographic archives notes, the photograph is a kind of 'incomplete utterance' because it is 'a message that depends on some external matrix of conditions and presuppositions for its readability' (1982: 85). Furthermore, as he makes clear, the meanings it can generate in each culture or period 'harnesses the photographs to various representational tasks' (87).

This rebuilding process starts by considering what the photograph's trace of a gendered, labouring body – the working-class woman – signified in the mid-Victorian period. A large body of feminist Victorian scholarship notes how markers of femininity functioned as a central definer for a bourgeois identity often referred to as the cult of domesticity and the 'separation of spheres' (Davidoff et al. 1979; Davidoff and Hall 1983, 1987; Hall 1979, 1981). Such discourses on gender were being established at the very moment Munby started to collect his photographs. However, other class-based signs contradicted this bourgeois norm to create a quite different conception of femininity – working-class femininity – and it was within these emerging and often contradictory signs that these photographs took on a commercial value.

Victorian scholarship on space identifies how the masculine, outside world was increasingly constructed as immoral, impersonal and dangerous, in opposition to the protected, sacred space of the home and femininity. Because working-class women worked in the outside space, whether rural or urban, they became tainted by these popular cultural definitions of immorality. Such readings increased as official discourses sought to 'study' sexuality, as identified by Foucault (1981). In this setting, the working classes were often imagined as sexually and morally uncivilized. In an urban context, this was exemplified by middle-class understandings of working-class women as prostitutes: 'the category of prostitute was not fixed or internally coherent; it was accommodating and flexible and could define any woman who transgressed the bourgeois code of morality' (Nead, 1988: 94; also see Walkowitz, 1980). These wider discourses on gender and location can be employed to begin to explain why Elizabeth Fairhurst did not want her photograph displayed for all to see in the shop window.

Nonetheless, Munby's records indicate that by 1865 the production and sale of photographs of pit brow women in their dirty working clothes had become a roaring trade; there were at least five photographers in the Wigan area. In 1863, he notes, he:

> went out soon after nine, to Dugdale the photographer ... found he had lately taken a good picture of one Jane Horton, aged 19, a collier at Patricroft (Ince Hall) ... saw this and one or two other 'carte de visites' of girls in pit clothes hung up in the street by the photographer'.
>
> (19/8/1863)

By 1865:

> ... to Dugdale, who has at length taken a group of pit wenches ... to Copper, who sells beer as well as photos, and who said he had sold hundreds of carte de visites of the collier girls ... and to Little ... a case of photographs hung on view in main street, and among them were several portraits of pit girls in costume ... All four said they sold these photos chiefly to commercial travellers'.
>
> (17/3/1865)

Contradiction and overdetermination: the slippery meanings of photographs

Why these women had been selected out for particular discursive representation can be understood through a recall of studies on the industry. This part of the mining industry was located in the north of England, and while this was not an urban-based form of labour as such, coal held a symbolic connection to urbanization, allowing it to function in a modern working-class context. A preceding 'fascination' in the labour of northern miners allowed them to be quickly positioned as emblematic of a certain type of English working-class labour.

Historical scholarship on the mining industry in England identifies how these women were marked out as a 'curious' type of female labour in earlier tourist literature. Angela John (1984) quotes Richard Ayton's *Voyage Round Great Britain* (1813), 'ragged and beastly in their appearance and with a shameless indecency in their behaviour' (John, 1984: 31). By 1840, this cultural positioning of the women's labour as 'indecent' collided with the rising definitions of bourgeois femininity, leading the Children's Employment

Investigation of the Mining Industry of 1840 to include women, although they were outside their remit (see John, 1984). Extracts from the report, including its use of woodcuts as evidence, were reproductions in the newly formed sensationalist press (Beegan, 2008). *The Morning Chronicle*, *The Penny Chronicle* and the *Leeds Mercury* reproduced extracts emphasizing 'the moral implications' (John, 1984: 45), and drew from terminology used to describe working-class prostitution: 'no brothel can beat it' (John, 1984: 40; see also Humphries, 1981).

By the 1840s, these women's labours had begun to acquire a popular cultural meaning of 'immorality'. In 1846, when the London and North Western Railway company opened up the North for popular internal travel, the women were subjected to a wider tourist gaze. This pre-photographic history can explain why mining women in their dirty working dress appears to have been one of the very first commercial uses of photography, in which these photographs are a precursor to the postcard. This is because these pre-existing meanings could activate the signs 'captured' in the photographs to trigger class and gender-based sexualized meanings immediately.

Turning attention back to the photographs, the issue of costume/fashion becomes central: the Wigan women had been instructed to come and be photographed in their dirty working clothes that included the wearing of men's trousers or breeches. Wider research on how dress and fashion functioned in this period identifies it as possibly the most important visual marker of class difference. Breward's (1999) study, drawing upon Dawson (1991), identifies how fashion for the middle classes was a conscious part of identity creation, 'something that has been "made up" in the positive sense of active creation' (Dawson, 1991 quoted in Breward, 1999: 14). Studies have revealed the co-dependent relationship between what one wore and its discursive meanings, confirming how the 'fashion ideals promoted by the clothing trade' soon began to share 'a common language with other more imaginative texts, in which the identification and assessment of urban "types" through visual codings was of central importance' (Breward 1999: 55). In class terms, fashion became a key visual signifier in establishing the class position of self and others.

The mining women had a quite specific working costume that marked them out from other working-class women: they wore working men's trouser-breeches, often accompanied by a man's jacket, and this made them signify masculinity more forcefully than other female trades. It is this that has been caught by the camera.

Moreover, if we return to Linkman's more detailed reading of gender, class and pose, it becomes apparent that other signs in the photographs also signified masculinity. The women are standing facing the camera, their feet, in large working boots, are apart, arms on hips. Furthermore, the enterprising photographers also included props to further signify their tough manual work. They gave the women a riddle used for sorting the coal and a large shovel to hold in the photographs (Figure 4.1). On a few

Figures 4.1 and 4.2 Commercial carte de visites by R. Little, Wigan 1867. Reproduced with kind permission from the Master and Fellows of Trinity College Cambridge.

occasions, this is accentuated with a large piece of coal and even a large wheelbarrow (Figure 4.2). Such props were not just employed to identify these women from other types of working-class occupations, but also imaginatively imported the suggestion of activity and labour into the stillness of the photograph. In a few cases, this was even more forcefully signified when the photograph had been taken outside, suggesting that the women had been photographed at their actual work.

It is also worth considering what the women's dirt signified in the photographs. The signs of her dirt and dark occupation again placed her in stark opposition to the light, clean and morally superior space of femininity in the bourgeois home, again importing wider discourse of uncivilized immorality while also drawing upon emerging imperialist discourse on racial difference.

The point to make is that the earliest commercially made photographic representations of a particular 'type' of working-class woman succeeded because they drew upon already circulating meanings that put the iconic codes held in the photograph to work. At the same time, the new indexical action of the photograph, with its myth of objectivity, could verify such ideologically motivated discourse as 'truthful'.

Nonetheless, in this first period, such an interpretation of the photograph as containing a fixed and truthful message was not yet fully secured. In 1863, at the conference of the National Association of Coal, Lime and Ironstone

Miners of Great Britain, the miners aired their objections to women working in the mining industry, and, in 1864, 14,000 miners signed a petition to the government containing the following complaint: 'the practice of employing females on or about the pit bank of mines and collieries is degrading to the sex, leads to gross immorality and stands as a foul blot on the civilization and humanity of the kingdom' (cited in Report from the Select Committee on Mines, 1867, Appendix, iii). In response, the government established a male committee to hear evidence in the House of Lords in London during 1865, 1866 and 1867. One task was to examine the validity of this claim, to adjudicate if the women should be banned from mining work. The recordings contain no testimony from the women miners, but called miners and mine owners from different locations: Shropshire, South Wales and Lancashire. Each was asked a specific question related to occupation, dress and immorality.

On April 15 1866, Munby recorded that the agents for the coal owners informed him that they 'of course wish it to be retained [the women workers] & the men have been telling lies against it' (cited in Hudson, 1972: 223). On June 14 1867, before the end of the sitting, the MP, Charles Neate, a member of the committee, informed Munby that they would not be interfering with female labour (226). It was in this disputed environment that photographs of pit brow women took on a new role as official evidence.

The committee questioned the miners on the women's dress: 'the entire person of the woman is covered and there is nothing indecent in the dress ... you spoke of the dress as being one of the leading features of the degrading character of the employment' (Minutes of Evidence taken before Select Committee of Mines. PP.XIV, 20). A Mr P. Dickinson, from the Wigan Coal and Iron Company, noted that it was 'rather a man's dress; and I believe, in some cases, it drowns all sense of decency betwixt men and women, they resemble one another so much ... it clothes the person, but it does not drown the feeling' (20). This reading of dress was echoed in the evidence given by other miners. However, what is most revealing in terms of the new invention of photography is that the committee used commercially sold photographs to evidence the miners' claims. Mr W. Pickard, the miners' agent of the Wigan district, a working miner for sixteen years, was passed a photograph:

> This is a photograph taken of a woman employed on the pit brow of a mine in the Wigan field; do you consider that a fair representation of the dress of the women so employed (handing a photograph to the witness)?
>
> Pickard replied, 'Yes; they are frequently seen in the town of Wigan in that complete dress, with a man's overcoat and that costume'.
>
> (64)

However, to contradict this visual evidence, Mr Gilroy, the principal manager of Ince Hall Coal and Cannel Company, produced his own photographic evidence:

> 'I think you have some photographs that you can show to the committee, which will give you some idea of the appearance of the girls employed on the pit bank of the collieries?' – I have. (The Witness handed a photographic album to the Committee.)
>
> (ibid: 363)

Gilroy, it appears, had sent the women in his employment to Cooper's, the photographer, to have them photographed in their Sunday best (John, 1984: 184). He anchored the meanings of these photographs by explaining how the 'women attended Sunday school and bible classes' (Select Committee of Mines. PP.XIV, 363).

While I cannot be exactly certain what the photographs looked like, I would surmise that they followed the conventions of early working-class domestic photography, aptly called the 'Sunday trade' (Edge, 2008) because working-class people started to have their 'likeness' taken on their day off, dressed in their Sunday best. The Munby archive holds a number of photographs of mining women in their Sunday best from this period, and I would be confident in suggesting that these are representative of the types of photographs submitted by Gilroy.

In these photographs, the iconic signs of a woman, dress, pose, props and location are quite different from the ones of her in working dress. Her dress now covers her legs, signifying cheap but respectable working class femininity, and her pose of sitting rather than standing signals frail femininity. Her arms are now crossed with hands resting on the table or her lap, signifying passivity rather than activity. Her restrained hair and clean, washed hands and face speak to her aspiration of becoming a 'civilized' member of society. The exchanging of props from coal, shovel and riddle to her bible can now signify passive religious respectability and, finally, the substitution of her trousers for a dress displaces the just too overt meaning of masculinity.

In a few cases, Munby collected photographs of the same woman in each of these guises – working attire and Sunday best – mirroring what was presented to the committee. In Figure 4.3, Munby has noted on the back of the carte de visite that this is 'Jaan Brahm Bottom Place Pit 1867', and that the other (Figure 4.4) 'is also Jaan Brahm 1867'.

The cultural 'messages' held in each of the photographs have been mobilized to support different cultural/political positions. However, in this context, they also somewhat ironically revealed the unsettled nature of photographic truth in itself. Because the photographs were of the same woman, who now signified almost diametrically opposed meanings, this first use highlights that a photograph can never actually record a 'truth' about the subject represented.

What this example illustrates is an important academic hypothesis on identity and identity formation: that class and gender are not the result of biology or an essential characteristic of any group of people, but are always discursive. In other words, such differences never reside in the body of a human

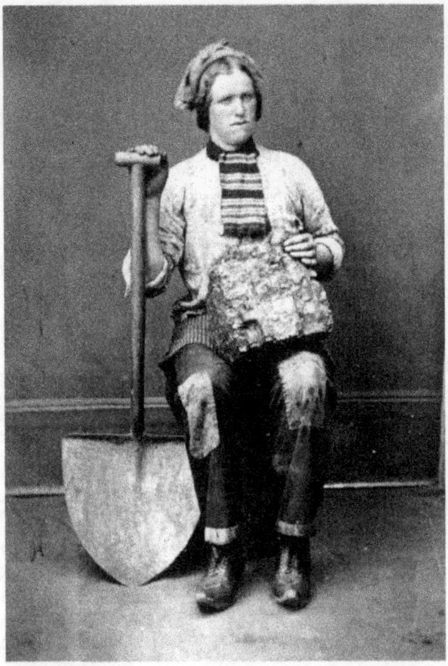

Figures 4.3 and 4.4 Jane Brown, Bottom Place Pit 1867. Reproduced with kind permission from the Master and Fellows of Trinity College Cambridge.

subject but are fabricated in popular cultural discourses. In terms of gender, each of the photographs was made via a process of tracing the corporeal body of a woman moved into representation through its transformation into an iconic sign system. In academic terms, we could term this iconic sign as her sex, the female body without cultural meaning (Butler, 1999). On to this passive sign, connotations go to work that are activated by the multiple discourses of gender circulating in popular culture. In other words, what this historical example remarkably exposes is that femininity is fabricated, and that it can shift and change in different historical environments, contexts and locations.

Class identity is perhaps more elusive as it has no originating sign in the body (there is no essential difference between the working-class or middle-class body). Instead, it as Thompson has noted: an identity 'largely determined by the productive relations into which men are born – or enter involuntarily' (Thompson, 1980: 9). In this historical moment, as I have revealed, these photographs were also heavily implicated in the discursive construction of class in which one image could connote an almost 'pornographic', aberrant feminine/working-classness, the other a respectable feminine/working-classness. Again, across the two photographs, the same woman signifies two different kinds of class-based femininity.

To shift this historical example into a contemporary context, we can borrow from Judith Williamson's (1986) reading of the postmodern photographic work of Cindy Sherman (*Untitled Film Stills* Series 1977–1981) when she argues that what:

> is crucial ... is the opposition between the images. 'Essentially feminine' as they all are, they are all different. This not only rules out the idea that anyone of them *is* essentially feminine but also shows, since each *seems* to be that there can be no such thing.
>
> (1986: 102–3)

Once these historically specific meanings have been rebuilt, they reveal how such a powerful discourse of difference impacted on the 'lived, cultural' experience of being a working-class mining woman in this period. The discourses activated by the cultural signs held in the two photographs of the same woman were mobilized by two different groups of men for their own interests. One 'spoke' (Figure 4.3) for the male miners as the hegemonic force (Gramsci: 1971) of patriarchal domestic ideology began to take hold, and they internalized aspirations for their own 'Angel in the House' (desires also articulated in the evidence given to the committee). The other 'spoke' (Figure 4.4) for the male mine owners, like Gilroy, who sought to keep these low-paid workers (women were paid the same as boys) as producers for their industry, as well as seeking to ensure that a different kind of low-class femininity was adequately constructed to secure a demarcation across different types of patriarchal power.

In this historical detective work, the 'voice' most difficult to excavate has been that of these working-class mining women. However, if we return to the photographs, we can perhaps discern an alternative reading: Wigan mining women pose in their hundreds for a self-image which can be read as a robust and hard-working person with a long tradition in the industry. Added to this, Munby indicates that the women personally owned such images, and in 1863, one of the photographers tells Munby that the women 'rather liked to be exhibited' (19/8/1863). The same women also recorded themselves at their 'feminine' best, taken in their Sunday dresses. This limited information allows us to at least speculate that in this transitional moment there may have been no contradiction across the two discourses, which together constituted her identity as a working-class woman.

To conclude, research that uncovers the varied and often disputed meanings of historical photographs should not been interpreted as confirming that photographs are, as Barthes warns, somehow semiotically 'anarchic'; that is, that the reader can read the image in whatever way they want because, as revealed here, readings of photographs are always created by the reader's cultural appreciation of the iconic signs in a photograph, and these are based 'on the different kinds of knowledge – practical, national, cultural, aesthetic' that different groups have 'invested in the image' (Barthes, 1982: 46).

Acknowledgements

Photographs reproduced with kind permission from: The Master and Fellows of Trinity College Cambridge.

References

Anderson, Benedict (1991) *Imagined Communities*. London, Verso.

Barthes, Roland (1973) *Mythologies*. London, Granada.

Barthes, Roland (1982) *Image Music Text*. Translated Stephen Heath. London, Fontana.

Barthes, Roland (1984) *Camera Lucida*. London, Fontana.

Beegan, Gerry (2008) *The Mass Image: A Social History of Photomechanical Reproduction in Victorian London*. London, Palgrave Macmillan.

Breward, Christopher (1999) *The Hidden Consumer: Masculinities, Fashion and City Life, 1860–1914*. Manchester, Manchester University Press.

Burgin, Victor (1986) 'Re-reading Camera Lucida' in Burgin, Victor (ed.) *The End of Art Theory: Criticism and Postmodernity*. London, Macmillan, pp. 71–92.

Butler, Judith (1999) *Gender Trouble: Feminism and the Subversion of Identity*, 10th anniversary edn. New York, Routledge.

Davidoff, Leonore and Hall, Catherine (1983) 'The Architecture of Public and Private Life: English Middle-Class Society in a Provincial Town, 1780–1850' in Fraser, D. and Sutcliffe, A. (eds). *The Pursuit of Urban History*. London, Edward Arnold, pp. 326–347.

Davidoff, Leonore and Hall, Catherine (1987) *Family Fortunes: Men and Women of the English Middle-Class 1780–1850*. London, Hutchinson.

Davidoff, Leonore, L'Esperance, Jean and Newby, Howard (1979) 'Landscapes with Figures: Home and Community in English Society' in Mitchell, J. and Oakley, A. (eds). *The Rights and Wrongs of Women*. Harmondsworth, Penguin, pp. 139–176.

Dawson, Graham (1991) 'The Blonde Bedouin' in Roper, M and Tosh, J. *Manful Assertions: Masculinities in Britain since 1800*. London, Routledge, pp. 113–144.

Edge, Sarah (1998) 'The Power to Fix the Gaze: Gender and Class in Victorian Photographs of Pit-Brow Women'. *Visual Sociology*, Vol. 12 No 2, 37–56.

Edge, Sarah (2008) 'Urbanization: Discourse Class Gender in Mid-Victorian Photographs of Maids – Reading the Archive of Arthur J. Munby (1828–1910)' *Critical Discourse Studies*, Vol. 5, No.4, 303–317.

Edge, Sarah (2013) 'Photography and Poststructuralism: The indexical and Iconic Sign System' in Dillet, B., Iain MacKenzie, Robert Porter (eds) *The Edinburgh Companion to Poststructuralism* Edinburgh, Edinburgh University Press, pp. 311–333.

Edge, Sarah (2016) *The Extraordinary Archive of Arthur J. Munby: Photographing Class and Gender in the Nineteenth Century*. London, I.B. Tauris.

Edwards, Elizabeth (ed.) (1992) *Anthropology and Photography, 1860–1920*. New Haven, CT, Yale University Press.

Foucault, Michel (1973) *The Birth of the Clinic*. London, Tavistock.

Foucault, Michel (1979) *Discipline & Punish: The Birth of the Prison*. Harmondsworth, Penguin Books.

Foucault, Michel (1981) *The History of Sexuality, Volume I: An Introduction*. Translated Robert Hurley, Harmondsworth, Pelican.

Foucault, Michel (1986) *The History of Sexuality, Volume II: The Use of Pleasure*. Translated Robert Hurley, Harmondsworth, Viking.

Gramsci, Antonio (1971) *Selections from the Prison Notebooks*. Translated Quintin Hoare and Geoffrey Smith-Nowell. London, Lawrence & Wishart.

Great Britain, House of Commons, Select Committee on Mines (1867) Index to the Report from the Select Committee on Mines. London, HMSO (496, 496–1).

Green, David (1984) 'Classified Subjects'. *Ten/8*, 14, 30–37.

Green-Lewis, Jennifer (1996) *Framing the Victorians: Photography and the Culture of Realism*. Ithaca, NY and London, Cornell University Press.

Hall, Catherine (1979) 'The Early Formation of Victorian Domestic Ideology' in Burman, S. (ed.) *Fit Work for Women*. London, Croom Helm, pp. 15–32.

Hall, Catherine (1981) 'Gender Divisions and Class Formations in the Birmingham Middleclass, 1780–1850' in Samuel, R. (ed.) *People's History and Socialist Theory*. London and New York, Routledge and Kegan Paul, pp. 164–176.

Heathcote, Bernard and Heathcote, Pauline (1979) Richard Beard: an Ingenious and Enterprising Patentee. *History of Photography*, Vol.3, No.4, 313–329.

Hiley, Michael (1979) *Victorian Working Women. Portraits from Life*. London, Fraser.

Hirsch, Julia (1981) *Family Photography: Context, Meaning and Effect*. New York, Oxford University Press.

Holland, Patricia (2000) 'Sweet it is to Scan ... Personal Photographs and Popular Photography' in Wells, L. (ed.) *Photography: A Critical Introduction*. London and New York, Routledge, pp. 113–158.

Hudson, Derek (1972) *Munby, Man of Two Worlds: The Life and Diaries of Arthur J. Munby, 1828–1910*. London, John Murray.

Humphries, Jane (1981) Protective Legislation, the Capitalist State, and Working Class Men: The Case of the 1842 Mines Regulation Act. *Feminist Review* Spring, pp. 1–33.

John, Angela (1984) *By the Sweat of their Brow: Women Workers at Victorian Coal Mines*. London, Routledge and Kegan Paul.

Linkman, Audrey (1993) *The Victorians: Photographic Portraits*. London, Tauris Parke Books.

Mavor, Carol (1996) *Pleasures Taken: Performances of Sexuality and Loss in Victorian Photographs*. London, I.B. Tauris.

McClintock, Anne (1995) *Imperial Leather: Race, Gender and Sexuality in the Colonial Contest*. London and New York, Routledge.

McGrath, Roberta (1984) 'Medical Police' *Ten/8* 14, pp. 13–18.

McGrath, Roberta (2002) *Seeing Her Sex: Medical Archives and The Female Body*. Manchester, Manchester University Press.

Nead, Lynda (1988) *Myths of Sexuality: Representations of Women in Victorian Britain*. Oxford, Blackwell.

Reay, Barry (2002) *Watching Hannah: Sexuality, Horror and Bodily De-formation in Victorian England*. London, Reaktion Books.

Ryan, James R. (1997) *Picturing Empire: Photography and the Visualisation of the British Empire*. London, Reaktion Books.

Sekula, Allan (1982) 'On the Invention of Photographic Meaning' in Burgin, Victor. (ed.) *Thinking Photography*. London, Macmillan, pp. 84–110.

Sekula, Allan (1992) 'The Archive and The Body' in Bolton, R. (ed.) *The Contest of Meaning*. Cambridge MA, MIT Press, pp. 343–389.

Smith, Lindsay (1998) *The Politics of Focus: Women Children and Nineteenth-Century Photography*. Manchester, Manchester University Press.

Spence, Jo and Holland Patricia (1991) (eds) *Family Snaps: The Meanings of Domestic Photography*. London, Virago.

Tagg, John (1988) *The Burden of Representation: Essays on Photographies and Histories*. London, Macmillan.

Thompson, E. P. (1980) *The Making of the English Working Class*. Harmondsworth, Penguin.

Walkowitz, Judith (1980) *Prostitution and Victorian Society: Women, Class, and the State*. Cambridge, Cambridge University Press.

Williams, Raymond (1981) *Keywords*. Glasgow, Fontana.

Williamson, Judith (1986) *Consuming Passions: The Dynamics of Popular Culture*. London, Marion Boyars Publishers Ltd.

Working Women in Victorian Britain 1850–1910. The diaries and Letters of Arthur J Munby (1828–1910) Hannah Cullwick (1833–1909) from Trinity College Cambridge (1993) Oxford: Adam Matthew Publications.

5 Inventing the Victorian boy

S.O. Beeton's *The Boy's Own Magazine*

Jochen Petzold

In England, children's literature – by which I mean a literature specifically produced for children – first became an important field of cultural production during the Victorian period. While books written for the entertainment of children started to appear in the 1750s, their number was still rather small. But by the middle of the nineteenth century, a number of writers had started publishing books that were either specifically written for children, or were often read by them: writers like Captain Frederick Marryat (in the 1830s and 1840s), William H. G. Kingston (from the 1840s), Robert Michael Ballantyne and Charles Kingsley (from the 1850s). However, while their books may have been popular with young readers, they were clearly too expensive to be actually bought by them with their own pocket money. Magazines, on the other hand, were much cheaper, some selling for as little as a penny an issue. Thus, during the nineteenth century, magazines became an extremely important – and popular – medium of recreational reading for children and young adults.

A range of factors led to a large-scale expansion of the magazine market in the second half of the nineteenth century: technological advancements meant that paper could be produced more cheaply and in better quality, which in turn meant that faster steam-powered printing presses could be used (cf. Altick 1957: 277). At the same time, the growing railway network meant that printed matter could be distributed quickly and comparatively cheaply. Furthermore, from mid-century onwards, the three "taxes on knowledge" were abolished: the advertising tax in 1853, the stamp duty on newspapers in 1855 and the paper duty in 1861 (cf. Fraser 1981: 228). Particularly the repeal of the paper duty made the production of juvenile magazines considerably cheaper. Finally, the potential readership was growing, since as the century progressed, more and more working class children were able to read. Thus, a growing demand for reading materials coincided with decreasing costs of production, leading to an upsurge in the number of juvenile magazines published in the second half of the nineteenth century. Marjory Lang (1980: 17) claims that "over one hundred new titles appeared in the 1860s and 1870s", and Sheila Egoff (1951: 26) underscores their importance, suggesting that the "children's magazines of the nineteenth century were institutions as those of today can never hope [...] to be". In the context of cultural studies, juvenile magazines are particularly interesting objects for investigation, since their appeal to a mass-market and their high frequency of publication make them "particularly sensitive seismographs of taste" (Drotner 1988: 4).

In this chapter, I want to examine in detail one of the "turning points" in the development of juvenile magazines, the advent of Samuel Orchart Beeton's *The Boy's Own Magazine* in 1855. According to Kirsten Drotner (1988: 66–7), the magazine:

> initiated a trend that was to grow into a norm – namely, the grading of potential readers by age and sex. By the same token, the magazine marked the beginning of a change in juvenile papers from religious didacticism or secular rationalism toward moral entertainment where an extrovert, imperial manliness mattered more than introspective piety or dry memorizing.

If Beeton initiated this "grading of potential readers by age and sex", it can be argued that he "invented" the Victorian boy, at least in the sense of a target audience: by marketing a magazine specifically for boys, he must have made certain assumptions about his potential readership, and he made certain statements about those boy-readers through the kind of material that he published. The more so since Victorians believed that literature had the power to influence its young readers' moral development – hence the insistence on "healthy" reading voiced by many Victorian commentators, and their condemnation of so-called penny dreadfuls that were seen as a threat to morality and public safety. For example, Edward Salmon claimed in 1888 that most of the magazines "provided for the older boys and girls, are in every sense of the word, dreadfuls", suggesting that there was "no limit to their [...] pernicious influence" (184), and while supposedly unhealthy reading material may not have been as plentiful in the 1850s as it was deemed to be in the 1880s, the concept of what constituted healthy reading had not changed dramatically in those thirty years.

Of course, the relationship between 'boys', their reading matter and concepts of manliness have not escaped critical attention – Martin Green (1980: 3), for example, calls adventure tales "the energizing myth of English imperialism". However, much of this attention has been focussed on the last decades of the nineteenth century. For instance, Patrick Dunae (1980: 106) declares that an "examination of boys' literature and empire should begin with the 1870s, a decade regarded traditionally as the watershed of the new imperialism". I do not propose to question his assertion that the decade of the 1870s was "marked by a growing interest in British overseas activities" (*ibid.*) and that this can be traced in boys' papers of the time. However, his statement seems to imply that the empire played only a marginal role in earlier magazines. Indeed, Dunae (1980: 107) suggests that the "imperial idea remained inchoate in boys' literature until the closing decades of the nineteenth century". This claim has been contested and to some extent falsified by Christopher Mark Banham (2006) with regard to *Boys of England* (1866–99). In this chapter, I propose to look even further back, and I intend to do two things: I offer a fairly detailed analysis of *The Boy's Own Magazine*, which has received surprisingly little critical attention, with a specific focus on its first series (1855–62), arguing that during this period Beeton made small but significant changes to his magazine, developments that would be pushed even further by later editors of boys' magazines. Furthermore, I will analyse the concept of boyhood and manliness suggested by Beeton's publication. It will become apparent that many of the concepts that can be found in later magazines like *Boys of England* or *The Boy's Own Paper* can already be traced in *The Boy's Own Magazine*, and while it may not have been as outspoken about the empire as magazines in the last decades of the nineteenth century, it does not ignore the topic, but integrates it into its depiction of "manliness".

Developing an idea: the first series of Beeton's *The Boy's Own Magazine*

S.O. Beeton had already launched a successful magazine, *The Englishwoman's Domestic Magazine* (1852–1879), when he came up with an idea that was to prove highly influential for the market of juvenile magazines: a magazine clearly intended and marketed for boys only. The two existing biographies of Beeton and his wife say nothing about how this idea was developed. In retrospect, Beeton allegedly confessed that it had been his ambition "to be regarded as having founded a periodical literature for the young folk of our country, such as exists in no other land, sail you East or West" (Hyde 1951: 49), and in the preface to the first bound volume of *The Boy's Own Magazine* (1855) he stressed the importance of providing boys with high quality literature, implying that there was very little of this kind to be had (cf. Boyd 2003: 29). In a kind of mission statement published on the wrapper enclosing the first issues of *The Boy's Own Magazine*, Beeton had claimed that it could "safely be intrusted in the hands of all boys", where it would "do them good" – by giving them "healthful and moral excitement, calculated at once to produce pleasure and convey instruction", by arousing "feelings of ardent admiration for all that is good and noble", and by suggesting activities that will "enliven and strengthen the mental and physical powers" (Anon. [Beeton] 1855: n.p.). This insistence on *delectare et prodesse* (delight and instruct) is typical of juvenile literature of the time. In terms of the contents treated, Beeton is deliberately wide, with one clear exception: he deplored the fact that "feminine accomplishments" like knitting, crochet or embroidery "are found in almost every existing publication, and are to 'boys' entirely useless, if not distasteful"; consequently, *The Boy's Own Magazine* would "avoid naturally all these subjects" (*ibid.*).

In devising the general concept of the journal, Beeton's most immediate model had probably been the wave of "Peter Parley" books that had come to the British market in the 1840s – imported copies, pirated reprints or imitations of the highly successful series of American publications by Samuel Griswold Goodrich. In the analysis of Drotner (1988: 64), both genuine Peter Parley publications and their imitations "put facts into focus", and as we will see, this clearly influenced the choice of subject matter in *The Boy's Own Magazine*. In an article on "Lads' Literature, and Writers for Boys" that appeared in *The Boy's Own Magazine* in 1867, the anonymous author even admits that "Peter Parley gave the boys and girls a capital annual", but insists that it was Beeton who dared to attempt a periodical "that could be recognised as the fit and proper organ for the ventilation of all subjects of interest to boys" (Anon. 1867: 114).

The first monthly issue of the magazine came out in January 1855, and two pence bought you 32 octavo pages (i.e. the page was c. 19 cm high and 11 cm broad) with two columns of rather small print and eleven woodcut illustrations. As Sheila Egoff (1951: 15) suggests, by "modern standards *The Boy's Own Magazine* was poorly produced", and the content may be equally surprising to a modern reader, particularly if he or she has seen copies of one of the boys' magazines that came out towards the end of the century. For example, *The Boy's Own Paper* opened its first number (in January 1879) with an exciting school story, in which the young protagonist is instrumental in winning a rugby match. None of this excitement can be found in the first number of *The Boy's Own Magazine*: it opened with a six-page biography of Benjamin Franklin.

Biography was considered wholesome reading at the time, and religious magazines had frequently printed biographies of Christian martyrs and missionaries (cf. Drotner 1988: 98). In the first volume of Beeton's journal, this religious focus was substituted by what Drotner (1988: 98) calls "secular adventurers"; however, the biographies cannot be categorized quite so neatly, as they comprise a rather motley crew: politicians and military leaders, inventors, explorers and artists, stemming from various countries and centuries and representing a wide range of careers. Only in the first text is there a clear moral which encourages the readers to "follow the example" of Franklin in order to become "a happy and prosperous if not a great man" (Anon. [Wynne] 1855: 7), but clearly the biographies are offered as exemplary lives. While political leaders and explorers/inventors dominate the list, it is not exclusively pointing the boy-readers into the field of what can be widely described as public service.

In the first number, the biography of Franklin is followed by an "account of the cayman, and how to catch him" (Anon. [de la Gironière] 1855: 7); that is, by a mixture of factual information on an exotic animal and a hunting narrative. This article is followed by two unambiguously non-fictional pieces, both of which are the first parts of longer running series: "The Tools of War", running through four instalments (vol. 1, nos. 1–3 and 5), introduces a range of different weaponry; "Famous Places" is a six-part series that gives detailed descriptions of Naples (vol. 1.1), Pompeii (vol. 1.2), Florence (vol. 1.3), Rome (vol. 1.4), Paris (vol. 1.5) and London (vol. 1.6). On page 18, we find the first piece of genuine fiction, a shortened version of a tale by Edgar Allan Poe, "The Thousand and Second Tale [of Scheherazade]", which presents modern-day inventions and events as unbelievable wonders (explained in the footnotes). The editor seems to have felt a need to justify the inclusion of fiction, for he adds a final paragraph not to be found in Poe's original text: "And thus we see how common facts in nature and art are stranger even than the wonderful fictions of the 'Arabian Nights'" (Anon. [Poe] 1855: 24). In a similar vein, the "Ballad of Robin Hood" is preceded by a short introduction that points to its antiquity and insists on its educational value:

> If the study of Latin is likely to make a scholar expert in the use of polished language, the study of old English ballads is highly conductive to the acquisition of that best of all English – strong, simple Saxon-English. It is, therefore, not *only* for the amusement of our readers, but also as a study, and as an example of the common style language of an age long past, that we print the following.
>
> (Anon. 1855 "Ballad": 25)

The reference to the study of Latin can be seen as an indication for the intended audience of the magazine. According to Drotner (1988: 70), *The Boy's Own Magazine* had not only "inaugurated a grading of readers according to age and sex, but also a literary separation according to class", and she argues that "a whole group of lower-middle-class and working-class juveniles was excluded" by Beeton. Indeed, if two pence a month might still have put it in the buying range of working-class children, the content clearly marked it as a periodical that would be more attractive to boys from the middle class, boys who had a first-hand experience of learning Latin, who might later consider a career where "Short Hand Writing and Reporting" (vol. 4, nos. 10 and 11) might come in handy, who might benefit from "Hints for Candidates Preparing for the University Certificate, or Title of A.A." (vol. 4, nos. 2, 3 and 4) and who had money to buy the materials necessary to try out the experiments or build the model ships or engines suggested in the magazine (see below).

The ballad is followed by yet another piece of non-fiction, "A True History of the Gunpowder Plot", and the last two and a half pages of the first number are filled with "Sports and Pastimes" – short pieces describing experiments or games – and "Facts, Fancies, & Phenomena" – short pieces giving facts about natural objects or phenomena. Taken together, a certain pattern starts to emerge. In its first numbers, the magazine relies heavily on non-fictional writing which is intended to inform and inspire. Exotic animals are presented and narratives of hunting these animals are explicitly included. Weaponry is seen as a proper topic, as is political and military history. Fiction is clearly under-represented, but the little that is there is usually set in exotic locations.

In this abstracted form (and disregarding the dominance of non-fiction), the contents do not seem much different from later magazines for boys, and the topics may well have been of genuine interest to many boys. However, reading through the first volume leaves one with a strong impression that the style of writing does not really aim at younger readers. This is hardly surprising, since very little was actually written with young readers in mind, as very little (if anything) was written specifically for *The Boy's Own Magazine*. The biography of Franklin is clearly based on a much larger chapter in James Wynne's *Lives of Eminent Literary and Scientific Men of America* (New York 1850); the piece on the cayman, including both illustrations, is taken from Paul de la Gironière's *Twenty Years in the Philippines* (trans. Frederick Hardman, New York 1854); the descriptions of Naples and Vesuvius are taken from William Ware's *Sketches of European Capitals* (Boston 1851); Poe's tale first appeared in the USA in *Godey's Magazine and Lady's Book* for February 1845; the ballad of Robin Hood is traditional and has frequently appeared in print before 1855; and the history of the gunpowder plot is a shortened version of a text that had appeared (probably not for the first time) in vol. 10 of *The Mirror of Literature, Amusement, and Instruction* (London 1828). The only longer text in the first number for which I was unable to locate an earlier appearance in print is "The Tools of War", but given the overwhelming prevalence of reprinted material it would appear unlikely that this should have been originally written for *The Boy's Own Magazine*. Of the authors, only Poe is mentioned by name, but the fact that the story was actually written by Poe is not clearly stated. Thus, it would seem unlikely that Beeton had obtained permission to reprint any of these texts. However, Beeton's sources were first published in America, and since no copyright agreement existed between Britain and the USA, this form of piracy was not illegal – and quite common (cf. Ghodes 1944: 16–17).

Thus, it is clear that in preparing *The Boy's Own Magazine*, Beeton had not approached writers to produce texts with a youthful audience in mind; rather, he had looked for and used texts that he thought would do. In the twelve numbers of the first volume, the only longer text that definitely made its first appearance in the pages of *The Boy's Own Magazine* is Thomas Miller's animal-story "The Life of a Dog", which started in July 1855 and was kept up to the end of the year; incidentally, it is the only story for which the author is clearly identified in the paratext.

Given the sometimes inadequate style, it is difficult to assess how successful Beeton's *The Boy's Own Magazine* was in its first year. Its focus on a male readership was unique for a juvenile magazine, and the topics covered – biographies of 'great men', natural history, exotic animals, hunting, travel, military history, weapons technology – would, with some alterations, form the backbone of many boys' magazines to come, so it is safe to assume that they were attractive to the intended readership. Beeton himself claimed almost instant success. In the preface to the first bound edition to the full first volume, he speaks of the scepticism that confronted his plan, and then presents a success story:

> Well, a few months decided the question. Our readers, instead of remaining a few, as was predicted, increased by thousands, and went on increasing. We then pitched the tone of the Magazine a little higher and more Boys – more thousands of Boys – rushed to buy. The experiment succeeded.
>
> (quoted in Boyd 2003: 29)

Of course, Beeton would sing his own praise, and his biographer Montgomery Hyde (1951: 50) concurs: "as the lack of suitable reading for boys in attractive form had long been felt, it [the magazine] was an immediate success". However, Nancy Spain (1948: 86) is less sure of this success: the "subscription list was causing trouble and Sam was forced to offer prizes to stimulate interest in the first number". Spain's accuracy with regard to *The Boy's Own Magazine*, which she consistently calls *The Boy's Own Journal*, is somewhat questionable, but it is certainly true that Beeton offered "Twenty Silver Watches of the full value of five guineas each" (Anon. [Beeton] 1855: n.p.). In order to enter the prize drawing, one had to send in a full set of slips that had to be cut off the wrapping of each monthly number, and this lottery scheme may well have generated interest in the publication. However, this method was not unusual, and Beeton himself had previously used it for launching his *Englishwoman's Domestic Magazine*, which was clearly successful. Whatever the exact sales figures over the first numbers may have been, "Beeton had proved that a market did exist for the boys' magazine" (Boyd 2003: 29). He soon found imitators: *The Boy's Own Journal and Youth's Miscellany* was launched in 1856, only to be bought up and incorporated into *The Boy's Own Magazine* in 1857, underscoring the success of Beeton's publication (cf. Egoff 1951: 33).

This growing success may well have been helped by small alterations Beeton made during the first years of the magazine. The general format and the size and price were retained from 1855 to 1862 (only the number of pages was raised from 32 to 42 in 1860), but the content did change perceptibly, if not radically. As Margaret Dalziel (1957: 76) has pointed out with regard to popular reading in mid-century more generally, "people wanted fiction"; and the most important general observation about the development of *The Boy's Own Magazine* is that fiction gained in importance. In the first volume, fiction was clearly marginalized and many numbers included only one or two short fictional pieces, none of which were used as the starting piece on the front page of an issue. The first number of the second volume (January 1856) then opened with "Elephant Stories", and while these stories do contain factual information, they are also exciting, action-packed and set in exotic places. They were followed by a story about a seaman in the American navy, a ghost story, a story in which the protagonist is granted 1,000 years of life, and finally a short fictional piece on "A Pet Rat". Of course, the issue also included non-fiction: a biography of Carlo Buonaparte, two short articles on bamboo and on sleepwalking, and factual information on horses and birds, but fiction had clearly gained much ground. This trend was continued in the third volume (1857), which opened with "Pym's Narrative", an unattributed abridgement of the first thirteen chapters of Edgar Allan Poe's *Narrative of Arthur Gordon Pym of Nantucket* (1838). Although not written for children, this story is particularly interesting because it uses a boy of around eighteen as the main protagonist. Thus, for the first time, the magazine offered its readers a boy-adventurer. The fourth volume (1858) then started each number with a piece of fiction, usually with an instalment of "The Real Robinson Crusoe", a story about Alexander Selkirk that was in fact a shortened version of *The Solitary of Juan Fernandez; or, The Real Robinson Crusoe*, Anne Wilbur's English translation of Joseph-Xavier Boniface's *Seul* that had first appeared in Boston in 1851. Furthermore, in 1858, *The Boy's Own Magazine* first printed translations of stories by the German author Friedrich Gerstäcker, actually naming the author; Gerstäcker would also feature prominently in the fifth, seventh and eighth volumes (1859, 1861, 1862). Hence, a clear pattern emerges: over the course of the first series (1855–62), fiction not only became comparatively more important, but there is also a trend towards longer narratives serialized over a few months or even the whole year, and there is a marked increase in original fiction published in the magazine. Most of the fiction printed in *The Boy's Own Magazine* can be broadly termed "adventure fiction", and most of it is not set in England, but either on board of ships or in foreign countries like the USA, Australia, India or on the African continent; historical romances set in the Middle Ages were also frequently included.

While fiction gained in importance, non-fiction still played an important part and commanded more space than was usual at the time in Sunday school magazines (cf. Drotner 1988: 101). However, changes can also be discerned in this field. For one, the strong reliance on biography was reduced: while each number of the first volume opened with a piece of biography, the second volume only featured six biographical essays. With one exception, the following volumes then include little or no biography; only the seventh volume (1861) features a seven-part biography of George Washington. Pieces on natural history remain a staple ingredient of *The Boy's Own Magazine*, and particularly in the later volumes of the first series they are frequently presented as longer series. For example, "The World of Insects" comes in seven instalments in 1857, eight in 1858 and another six in 1859, when it is superseded by eight articles on "The Marvels of the Microscope" and four on "Mysteries of the Crucible" (1859 and 1860). All of these series were signed "M.S.R.", suggesting that they were originally produced for the magazine, but I have been unable to identify the author. History, particularly military or naval history, was also regularly featured, but the topic was treated more regularly in the form of fiction. A topic that saw comparatively little treatment in early issues is that of physical exercises. This changed in 1859, when a series on “Manly Exercises” was started that was continued up to the end of the first series (1862) to a total of thirty instalments, offering advice on fencing with a rapier (1859), rifle shooting (1860), fencing with broadsword and singlestick (1861), gymnastics (1861 and 1862) and general “training” (1862).

These essays are also indicative of yet another development: the increasing attempt to have a direct, active influence on the readers’ lives. Participatory elements like a correspondence page, essay competitions and a puzzles page were regular features of the magazine from the very start. To these were added articles that offered advice on various pastimes and hobbies: the selection and management of domestic animals, or the construction on various models or machines. Furthermore, many of the non-fictional series included detailed instructions for experiments.

On the whole, over the course of the first series, Beeton greatly expanded the role of fiction, and he widened the range of non-fictional topics. This combination of ingredients was clearly attractive to boys and in the early 1860s, at the end of its first series, the magazine reputedly sold some 40,000 copies per month (cf. Hyde 1951: 53, Egoff 1951: 15). Possibly prompted by this success, Beeton started a “New Series” in 1863 and greatly expanded the number of pages to 92; accordingly, the price rose to six pence per monthly issue. The general structure of the magazine was retained, but the relative importance of fiction was further increased and numbers now started with a piece of serialized fiction, usually a historical romance. While authors were very rarely named in the first series, this now became the norm, indicating that most pieces were now either specifically commissioned for *The Boy’s Own Magazine*, or at least appeared there for the first time. This also applied to the non-fiction: for example, the Rev. John George Wood, already a well-known author on natural history, regularly contributed to the magazine. The magazine now took pride in its writers, even publishing two plates with their portraits (NS vol. 9, nos. 50 & 51, 1867), and even H. John Allingham (publishing under the pseudonym Ralph Rollington), who was clearly no friend of Beeton’s, conceded that “had the level of excellence reached in the years 1863–66 been preserved, the magazine might still have been a flourishing and popular book for boys” (Rollington 1913: 20). There is no space here to examine to what extent Allingham’s accusation of deteriorating quality after 1866 is justified; what is clear, however, is that after the failure of the banking house Overend, Gurney and Co. in May 1866, Beeton was forced to sell the copyright of his successful magazines to the publishing house Ward, Lock and Tyler to escape personal bankruptcy (cf. Hyde 1951: 133–4). He remained editor of *The Boy’s Own Magazine*, but business cannot have been good, for in 1870 a second “New Series” was launched, which was not in fact “new’, but reprinted previous issues, starting with the first “New Series” of 1863.

In 1874, the *Boy's Own Magazine* disappeared off the shelves, after almost twenty years of publication. In terms of popularity, it was soon outrun by weekly papers like Edwin J. Brett's sensational *Boys of England* (1866–99), which quickly sold 150,000 copies, with sales rising to c. 250,000 in the 1870s (cf. Banham 2006: 1); or the Religious Tract Society's more respectful response, *The Boy's Own Paper* (1879–1967), selling some 200,000 copies in its first years, the numbers allegedly rising to over 500,000 in the 1880s (cf. Dunae 1976: 133) – although recent scholarship suggests that sales were considerably lower (cf. Banham 2006: 32; McAleer 1992: 217–9). However, by publishing *The Boy's Own Magazine*, Beeton had created a pattern that all later boys' magazines were to follow, at least to some extent; he had, in fact, invented the boys' magazine. In the following section, I will examine to what extent he also invented the Victorian boy.

Constructing "boyhood" and "manliness" through *The Boy's Own Magazine*

As we have seen, *The Boy's Own Magazine* offered a wide range of topics, and it is hardly possible to reduce its concept of an ideal boy to a single idea. However, certain patterns do emerge – patterns that will also be found in later publications for boys. Most important is the 'can do' spirit that dominates both fiction and non-fiction. Protagonists are active and successful, and they achieve success by fair means. These may well be violent – innumerable stories, particularly the historical romances, include descriptions of military encounters – but cruelty is criticized. The magazine's politics in this regard become clear early on. The first volume (1855) includes a piece on "War Scenes", with explicit reference to the "present war against Russia" (i.e., the Crimean War, 1853–1856), which holds important implication for the concept of manliness. The essay starts with the following observation: "War is not only inevitable, but sometimes just and righteous. It is dreadful to slay; but a battle-field is a less shocking spectacle than a nation dishonoured, craven, and meekly submitting to tyranny" (Anon. 1855 "War": 284). National honour is presented as an important value, and the author draws direct conclusions for the reader: he expresses his hope that "no boy of ours has been led into the belief that honour *is* nothing [...] and money everything" (*ibid.*). However, the author also cautions that violence is no end in itself: "Let's have no fighting for fighting's sake", he declares (*ibid.*), and the following descriptions of war scenes are not marked by bravado, but rather insist on the horror of war.

That the readers of *The Boy's Own Magazine* took this sentiment to heart is indicated by an essay competition in the fourth volume (1858). The topic was "True Courage", and the editor was "delighted to see that we have as our readers boys who are able to master the difficulties of composition; and, what is much better, thoroughly to understand the difference between virtue and vice, between the real and the spurious" (Anon 1858: n.p.). The topic inspired a comparatively large number of entries, sixty-one entrants are explicitly mentioned with short comments. It is probably safe to assume that the winning essay was selected because it reflected the politics of the editor; it certainly mirrors the sentiments expressed in "War Scenes". Courage, the winning essay points out, "is unquestionably one of the noblest attributes of humanity", and the author (signing himself "Rothennjoh") names courage in war as the most obvious example, claiming that the "history of our beloved country is as prolific of examples of true warlike courage as any" (Rothennjoh 1858: 268). However, the author insists that the "*power of endurance, the venture of life for the salvation of others, and obedience to duty under the most trying circumstances*, are much higher manifestations of true courage than that displayed in the field" (*ibid.*: 269, emphasis in the original). These sentiments are then exemplified with men and women who strove for human rights and with scientists and inventors, and the author concludes that the "peaceful triumphs of mind over matter we hold to be greater than those of man over man" (*ibid.*: 270).

Thus, the magazine offers a complex concept of manly courage that cannot be reduced to military heroism. At the same time, feats of British military bravery are proudly presented in numerous stories like "Battle of the Shannon and the Chesapeake" (vol. 3.1, 1857), and non-fictional essays like "British Arms in India" (vol. 2, nos. 6–12, 1856). Furthermore, weapons technology is amply examined in the first volume (1855) in the series "The Tools of War" (nos. 1–3 and 4). However, the military is not really presented as a career opportunity until a series of essays on various Cadet Corps is run in the first and second volume of the "New Series" (1863). The same volumes also give a detailed "Story of the British Navy", suggesting that military topics gain in importance. However, non-military courage remains an important issue throughout, and many stories exemplify courage as perseverance in seemingly hopeless situations (particularly after shipwreck), implicitly encouraging the readers not to give up too quickly.

The British Empire does not play a prominent role, but neither is it reduced to "a dramatic background for adventure", as Dunae (1980: 107) suggests with regard to the contemporary novels of Ballantyne and Kingston. As was already noted, imperial history is explicitly treated ("British Arms in India"), and the empire is repeatedly presented as a place where one might make one's fortune. Thus, we have stories about successful gold-diggers and farmers in Australia, who deliberately emigrated there from England (e.g. "A Digger in Difficulties", vol. 2, nos. 8 and 9, 1856, or "How I relished my first dinner in the desert wilds", vol. 3.5, 1857), and in "The Deadly Snake" (vol. 3, nos. 8–12, 1857), the narrator claims to have made his fortune in Ceylon. Furthermore, Beeton himself seems to have had the empire very much in mind in conducting his journal, if the following comment attributed to him is to be believed:

> No small thing I count it to help to form the taste and influence the mind of a youth whose glorious heritage it is to possess the empire that

> their fathers have founded or preserved, and whose duty it will be to hold that empire, handing it down greater, more prosperous, to future generations.
>
> (Anon. 1867: 113)

Since both fiction and non-fiction often concern foreign countries, the magazine also makes implicit and explicit statements about human "races". While Louis James (2009: 65) is right in pointing out that *The Boy's Own Magazine* "avoided the racist imperialism that was to dominate later boys' literature", the example he cites – the series "Cortes and the Conquest of Mexico" (vol. 1 nos. 2–12, 1855) – is only partially convincing. It is indeed noteworthy that the blood toll of the Spanish conquest of Mexico is called "inhuman and detestable", all the more so since the comment is not made in William Prescott's *History of the Conquest of Mexico*, on which the series is based. However, it is Spanish imperialism which is criticized, not British. While *The Boy's Own Magazine* is not aggressively racist, it certainly promotes the idea of a hierarchy of races. Thus, while slavery in the United States is deplored (some twenty years after the abolishing of slavery in British colonies), blacks are repeatedly referred to as "niggers", and the stories reprinted in the magazine show them to be intellectually inferior to whites. For example, "A Nigger 'Break-Down'" starts with the observation that if black people "where not whipped too much, and worked too much, and treated too much like brutes", they would be "the happiest people under the sun [...] always dancing, and singing, and clattering the bones, and riding, and making themselves boisterously happy without giving themselves the trouble of thinking about it" (Anon. [Peterson] 1855: 55) – combining criticism of US slavery with a belittling depiction of blacks as "naturally" thoughtlessly happy. This comment clearly expresses the belief of the editor (presumably Beeton himself), since the first paragraph is added to a story that is taken from the chapter "Rosa's Wedding" in Thornton Randolph's (really Charles Peterson's) *The Cabin and Parlor; or, Slaves and Masters* (Philadelphia, 1852).

Thus, the following image of the 'ideal boy' emerges from the pages of *The Boy's Own Paper*: he is interested in a wide range of topics (history, natural history, sciences; but also art and literature), as long as they are not thought of as feminine. He is no atheist, but neither is he overly devout or pious. He is physically strong (or striving to be) and shows his manliness by perseverance in the face of adversity, but also by moderation and chivalry. He is interested in all things connected to the military or the navy and is proud of his country, but does not like fighting for fighting's sake. He is against slavery, but assumes himself and his race to be superior to non-white races. All in all, he is not much different from the image created in *The Boy's Own Paper* some twenty years later – a strong indication that the active imperialism detected in magazines of the late-nineteenth century marks a change in politics that is one in degree, not in kind.

References

Only articles from *The Boy's Own Magazine* (abbreviated *BOM*) actually quoted are listed here; those merely referred to are documented by year and volume/number in the text. In the first series, many articles were anonymous reprints. The actual author is indicated in brackets if he/she could be ascertained.

Altick, R. (1957) *The English Common Reader*, Chicago & London: University of Chicago Press.

Anon. (1855) "A Ballad of Robin Hood", *BOM*, vol. 1 no. 1, pp. 25–28.

Anon. (1855) "War Scenes", *BOM*, vol. 1 no. 9, pp. 284–287.

Anon. (1858) "Prize Essay on 'True Courage'", *BOM*, vol. 4 no. 7, back wrapper, n.p.

Anon. (1867) "Lads' Literature, and Writers for Boys", *BOM*, NS vol. 9 no. 50, pp. 113–118.

Anon. [Beeton, S.] (1855) "Boys' Own Magazine", *BOM*, vol. 1 no. 1, front wrapper, n.p.

Anon. [de la Gironière, P.] (1855) "Catching a Cayman", *BOM*, vol. 1 no. 1, pp. 7–11.

Anon. [Peterson, C.] (1855) "A Nigger 'Break Down'", *BOM*, vol. 1 no 2, pp. 55–57.

Anon. [Poe, E.A.] (1855) "The Thousand and Second Tale", *BOM*, vol. 1 no. 1, pp. 18–24.

Anon. [Wynne, J.] (1855) "The Printer Boy", *BOM*, vol. 1 no. 1, pp. 1–7.

Banham, C. (2006) "*Boys of England* and Edwin J. Brett, 1866–99", PhD Thesis University of Leeds, http://etheses.whiterose.ac.uk/792/1/uk_bl_ethos_431946.pdf.

Boyd, K. (2003) *Manliness and the Boys' Story Paper in Britain: A Cultural History, 1855–1940*, Houndmills: Palgrave Macmillan.

Dalziel, M. (1957) *Popular Fiction 100 Years Ago: An Unexplored Tract of Literary History*, London: Cohen & West.

Drotner, K. (1988) *English Children and their Magazines, 1751–1945*, New Haven: Yale UP.

Dunae, P. (1976) "*Boy's Own Paper*: Origins and Editorial Policies", *The Private Library*, vol. 9.4, pp. 122–158.

Dunae, P. (1980) "Boys' Literature and the Idea of Empire, 1870–1914", *Victorian Studies*, vol. 24, pp. 105–121.

Egoff, S. (1951) *Children's Periodicals of the Nineteenth Century: A Survey and Bibliography*, Library Association Pamphlet No. 8, London: The Library Association.

Fraser, H. (1981) *The Coming of the Mass Market*, London: Macmillan.

Ghodes, C. (1944) *American Literature in Nineteenth Century England*, Carbondale, IL: Southern Illinois University Press.

Green, M. (1980) *Dreams of Adventure, Deeds of Empire*, London: Routledge & Kegan Paul.

Hyde, M. (1951) *Mr and Mrs Beeton*, London: George Harrap.

James, L. (2009) "Now Inhale the Gas: Interactive Readership in Two Victorian Boys Periodicals, 1855–1870", *Victorian Periodicals Review*, vol. 42.1, pp. 64–80.

Lang, M. (1980) "Childhood's Champions: Mid-Victorian Children's Periodicals and the Critics", *Victorian Periodicals Review*, vol. 13, pp. 17–31.

McAleer, J. (1992) *Popular Reading and Publishing in Britain, 1914–1950*, Oxford: Clarendon

Rothennjoh, A. (1858) "True Courage", *BOM*, vol. 4 no. 9, pp. 268–271.

Rollington, R. [i.e. H. John Allingham] (1913) "A Brief History of Boys' Journals", http://www.geocities.ws/justingilb/texts/Rollington.htm (viewed 8 September 2014).

Salmon, E. (1888) *Juvenile Literature As it Is*, London: Henry Drane.

Spain, N. (1948) *Mrs Beeton and Her Husband*, London: Collins.

6 Accept no substitutions!

Advertising, gender and 'race' in constructions of the consumer in the nineteenth century

Allison Cavanagh

Advertisements have an unusual status as cultural artefacts and historical documents. They are most often derided by cultural analysts, seen as intrusive ephemera which legitimate and extend the reach of a shoddy commodity culture and its underpinning ideologies. If popular entertainment is often wrongly dismissed, as Patrick Joyce one argued as 'the merest froth upon the surface of what really mattered' (1991: 216), advertisements are doubly so. Methodologically as well as conceptually, adverts are seen as a 'poor cousin' art form. It is argued that they do not embody or represent the cultural life of an era in the same sense as other cultural artefacts. Often anonymously produced, they are 'non-authored' works not originating from the unique vision accorded to the artist. Adverts are produced to order, in committee. On the other hand, it is argued they present extreme and magnified images of a society rather than its reality. They are seen as uniquely effective carriers of ideology. It is for good reason that the historian Michael Schudson (1984) argues that adverts' depiction of the present is sufficiently elaborate to justify a comparison with the ideology that imbues socialist art. Socialist realism in art required the artist to present positive images of the present, to show an optimistic view of the future, to use a simplified and accessible vocabulary of images, and show images of people worthy of emulation. In the same sense, Schudson argues, advertisements can be understood as 'capitalist realism', a 'set of aesthetic conventions... link[ed] to the political economy whose values they celebrate and promote' (1984: 214), for they too are diffused with optimism, display positive images of the 'typical' citizen as a role model, and use a range of accessible and easily to comprehend images. Thus he argues adverts can be understood as the typical form of art in contemporary capitalist society. Advertisements, then, are seen as lesser texts, both culturally and historically, presenting a view neither authentic nor artistic.

Whilst it is certainly true that adverts present an unrealistic vision of society, this is in a sense an unfair snub. The value of adverts to historians and cultural analysts does not reside in the extent to which they can provide a documentary image of the past, but in the sense that they offer a glimpse of the aspirations and fears of a society as these are encoded into texts where the chief requirement was easy readability. Of course in responding to this latter, adverts draw on a simplified and stereotypical lexicon of imagery, and the implications of this for the way adverts use gendered and racially inflected images will be discussed later on. Adverts do, however, offer us a unique glimpse into the ways social aspirations and concerns are mobilised and interpreted. As Lori Loeb has noted, adverts 'became a mirror and an instrument of the social ideal' (1994: 10). They also draw upon dominant social motifs and cultural imagery, in so doing, developing and extending the reach of the already popular. They, therefore, provide a useful lens in understanding the popular lexicons, styles and pre-occupations of an era.

This is especially the case in the nineteenth century. It is often observed of our own era that we are subjected to a constant bombardment of promotional texts, but our urban spaces are far less cluttered with advertisements than those of Victorian London and Paris. As the writer Victor Fournel's observed:

> Nowadays, Paris is nothing more than an immense wall of posters, studded from chimney to pavement with squares of paper of all colours and sizes ... a gigantic bazaar where advertising lies in wait for you on all the sidewalks, stops you, besieges you, hunts you down, snatches you by the eyes and ears, sneaks in your pocket, sits down in your theatre box, follows you home...
>
> (cited in Carter, 2012: 19)

Just as the physical spaces of the city were being deluged by advertisements, so the burgeoning of the popular press, fuelled by cheaper production methods, the reduction (1836) and abolition (1855) of stamp duty, speedy distribution through the railway network, and of course advertising revenue, provided another outlet. The competition in the mass press for advertisers became intense and advertisements took up large amounts of print space.

Advertising in the Victorian era, then, provided a vast body of cultural materials. However adverts, although trying to appeal to the widest possible audience, were far from a demotic art form. The consumer subjectivities that they constructed, along with core notions – for example community, respectability – that they appropriated, were not advantageous to all. Victorian advertising drew upon and reinforced profound inequalities, as we will see. I begin by looking at the ways in which particular styles of advertisements emerged in the Victorian era, before going on to examine how adverts constructed the notion of the consumer.

Tropes of advertising

Histories of advertisements seldom begin with the forms of the advertisements themselves. Partly for the reasons detailed above, analysts tend to approach them either as an offshoot of commercial history, or through locating them as an element of greater or lesser significance in the history of the press. Where advertisements are studied for the content, there is a tendency to reach down through the historical texts to seek a unity with the present, to look for the roots of present advert formats in those of the past. Of course such a unity is not an artefact of the scholar's view; many advertisement formats have been so successful that they have developed a life of their own, being reiterated and reinterpreted. An excellent example of this in the twentieth century is the 'soap opera' family story of advert; for example, the 'Bisto' family in the UK. Using the female body as an adjunct to car adverts is another, less homely but equally historically-rooted aesthetic. There is a tendency, then, to see advertisement imagery as somehow timeless and, therefore, logical, essential and natural. This apparent naturalness is of course an accomplishment, taken-for-grantedness disguising intellectual artifice. However, as many theorists have noted (see for example Roland Marchand (1985) in his analysis of 1930s advertising in America, and Leiss, Klein and Jhally (1997) in their work on nineteenth- and twentieth-century advertising), this supposed timelessness is in fact an illusion; different eras are dominated by distinct though multiple tropes of advertising. These can take the form of common stories, dominant and repeated images, 'parables' (Marchand, 1985) and set pieces that appear in advertisements. It is not a great leap, then, to see these as genres of advertisements. Such genres emerge out of the concerns and pressures of an era and as such, tell us much of the cultures which gave them birth.

A key pressure in Victorian advertisements, and one from which many tropes emerged, was what we might term a credibility gap and the need to escape public scepticism. The Victorian consumer had to be a 'wary beast' in an era in which consumer law was largely non-existent, and 'scams' and fraud commonplace. Advertisers, moreover, had long been dogged by public fears of 'puffery'. The historian Macaulay's 1830 diatribe against the puffing of books made clear the sense in which the practice was regarded as unacceptable by 'any man of honour and virtue' (Macaulay, 1830; cited in Nevett 1982: 111). 'Puffing' was one strategy resorted to, often unsuccessfully (Hindley and Hindley, 1972), to attempt to circumvent the taxes on advertisements in the early decades of the century; however, it was also a staple of the immensely influential patent medicine industry. The early patent medicine industry, as Twitchell (1996) has pointed out, devoted much of its profits to advertising. The need for such large investment is apparent, given that the products themselves had little or nothing to distinguish them from one another since these elixirs and potions contained little in the way of active ingredients. The testimonial then established the efficacy of the product, all the more so if it such endorsements could appear to have arisen spontaneously. Thus Dan Pinkham, son of the pill purveyor Lydia Pinkham, promised a large sum of cash to anyone who could prove one of the claims that accompanied his false. This had the advantage of encouraging a close reading of the texts, as much as acting as an expression of confidence (Twitchell, 1996).

Moreover, advertisements using this format simultaneously drew upon and constructed an intimacy and a sense of a community around the product. As Loeb has pointed out, adverts:

> draw on the unifying force of self-disclosure. Neighbours confide the secret for linen 'white as snow' over the garden wall. Best friends whisper to each other over tea. One woman confides the secret of beauty to another. The secret is an essential element of the Victorian advertisement. The advertiser cultivates the central idea of community-that when people self-disclose to each other, they grow together...The pleasing product is an agent of unity.
>
> (1994: 152)

This kind of unity was a necessary emollient when set against the backdrop of the social dislocation and mass urbanisation of the Victorian era. In this sense, then, the advert 'redefines community for the mass society' (Loeb, 1994: 157). People connected with each other rhetorically and symbolically in advertisements through the activity of consumption in the same sense in which, as Rappaport (2001) has noted, they connected physically in the new spaces of consumption through the activity of shopping.

The testimonial, then, creates a sense of mutuality and common purpose, and provides a 'solidity' to the product, being now something located in the social world of our known acquaintance. However, it was only effective to the extent that, first, as Dan Pinkham astutely realised, consumers were already engaged in a close reading of the advertisement (Twitchell, 1996) (something which, in the crowded landscape of Victorian promotions, was increasingly unlikely); and second, that the reader had confidence in the authenticity of the testimonial, something again in rather short supply. It is, then, not so great a step to move the advert viewer from their known circle to their imagined one, and thus Victorian advertisers turned to the image of the 'celebrity'.

The celebrity enters the lexicon of advertising styles, as Loeb points out, through the dominance of archetypes of the 'great man'. In an attempt to elevate products, advertisers sought to associate them with the already lauded in Victorian culture, and of these Loeb identifies the images of the scientist, royalty, the actress and the explorer as particularly prevalent. The explorer image was especially fruitful, feeding into imperialist and racially inflected narratives, as will be noted below. Celebrity adverting integrates these archetypes with the testimonial form, picturing known figures endorsing products. Thus one advert for Congo brand soap from 1891 carried a quote attributed to the explorer Henry Morton Stanley alongside a line drawing of the man himself carrying a box of the soap knapsack-style on a stick, presumably through the country for which the soap was named (Richards, 1990: 137). However, as Thomas Richards notes, this process of using celebrity to confer gravitas on a product was very much a 'bootstrapping' operation. The irony of advertising using supposedly charismatic individuals was 'that most of these figures were comparatively unknown before they appeared in the advertising columns and that the ones conferring celebrity were not the "celebrities" but the people publicizing them' (Richards, 1990:84).

Of course the most high profile of celebrity endorsers in the Victorian era was, unwittingly, Victoria herself. The image of the queen was used on an immense outpouring of manufactures and other goods, most notably, as Richards has pointed out, on the occasion of the Golden Jubilee in 1887. The Jubilee image of Victoria, he argues, 'unified and routinized the semiotics of commodity culture and created a lingua franca for late-Victorian advertisers' (1990:111). As a figure of the nation, she was classless; as a woman, she domesticated the commodity as much as elevated it. However, most significantly as a collector of goods (rather than a user or consumer per se), her image crystallised an emergent commodity system. Goods ceased to be simple rivals for each other. Instead, the commodity system emerged in which individual goods are rendered meaningful as part of a collection of objects. Advertising becomes central to this operation in producing the object as spectacle, Richards argues. As the collector par excellence and the single unifying face of the produced commodity, Victoria's image connected up narratives of nationalism, and gender (see below), and articulated them with commercial culture. Her image went beyond both product and brand, signifying the idea of consumption itself.

The need to establish credibility was, however, a constant pressure since it is subject to what might be termed semiotic exhaustion. The constant need to claim and defend credibility itself tends to undermine that claim, and the more products compete to demonstrate their validity, the weaker each claim appears. Thus Victorian advertisers used a range of other techniques and tropes. Key amongst these is attaching the commodity to the idea of luxury. As historians have pointed out (see Berry 1994, Smith 2002), the concept of luxury is historically highly mutable, both in the objects and practices to which it attaches and in the moral estimation accorded to it. To this day, the idea of luxury contains within it both the notion of a 'treat' and a pleasure, and a certain fraught negotiation over entitlement. In the Victorian era, this was no less the case than today. The invocation of luxury for an advertiser was a double-edged sword: on the one hand, inviting the audience to participate vicariously in supposedly elite practices through consumption; on the other hand, risking the appearance of 'putting on airs and graces'. The invocation of classical imagery smoothed the rougher horns of this dilemma. On the one hand, it tied in to the wider use of images of aristocracy as figures of emulation; on the other, it 'allowed the admirer to reveal an appreciation of higher ideals' (Loeb, 1994: 34). The imagery of luxury used was, then, one designed to appeal to a specific type of consumer, the new urban middle class, who after all possessed the greatest purchasing power for the new consumer goods. In this context though, the most important thing to note is that it inflected the image with an aura of 'timelessness', associating it with ancient virtues and dissociating from the shoddy rapidly produced consumables of nineteenth-century Britain. At the same time, as we will see later, such adverts, for example those who pictured Victorian women in the guise of the Greek graces or posed as classical statuary, could draw upon a broader sexual imagery.

A further technique, emerging naturally out of semiotics of luxury, is the lauding of the physical site of production. As McFall and Dodsworth (2009) have noted, architecture played an immensely important role in the promotion of life insurance. Insurance is a 'commodity' that is ephemeral in nature. The purchase of a policy does not buy something concrete and in any case, the policyholder hopes never to have to claim, most especially in the case of life insurance. Thus it is easy to appreciate the importance for insurance companies of translating that physical absence into a concrete presence in their promotional materials, providing the solidity that the product itself lacked and rendering it into a commodity, a thing in the world. The use of elaborate neoclassical architectural motifs, both in reality and in promotional materials served this aim, providing a physical site for this commodity at the same time connecting it with the theme of luxury.

A further trope to emerge is that of the mystification of the process of advertisement production itself. The advert came to appear as a natural, 'found' object rather than something produced by copywriters and ad men. One key technique was the mutation of the endorsement into the confessional. Confessional styles of advertising play on the idea that an endorsement comes about against the will of the endorser. The format, highly effective in health advertisements, is of a character confessing to a third party perhaps that they are sufferers of a particular condition and throwing in an endorsement as a by-blow. Another, more subtle variant of this is the use of the 'overheard' as a style, suggesting that the reader (of the advert) is not the intended recipient of the message. This format was effective for a wider variety of products, including those, like financial products, for which 'insider knowledge' might be said to be at a premium.

Finally, the credibility gap took another and darker form in Victorian advertisements: the constant assertion of originality and authenticity. Adverts routinely emphasised the idea that the product referred to was the 'original and only genuine', inviting the consumer to beware of imitation and piracy. Such invocations of course served to put the public on guard against the widespread practice of forgery in an era in which patents and copyrights were patchily and ineffectively enforced, if enforced at all. Equally, they sought to position the brand in question as the original, and therefore as the product other manufacturers aspired to copy, imitation being a sincere form of flattery (if a rather burdensome form of business). Moreover though, they served to rhetorically underscore a sense of the consumer as an imperilled innocent, needing the guiding hand of the advertiser to avoid the menace of shady practitioners.

The dominant tropes and narrative styles of advertisements, then, centred as they were on the notion of credibility, and the gap between the knowledge of the consumer and that of the advertisers created and reinforced an asymmetrical relationship between the advertisers and their audience. The audience is 'interpellated', in Althusser's sense, as an innocent (and potential victim) of an out-of-control commodity culture (see Althusser, 1970; in Storey, 2009). The advert appears as a guide, whether friendly as in the case of much 'celebrity' advertising and the 'overheard' narrative, or stern and didactic, and in so doing reorganises the relationships between the audience and the product, inserting a rhetorically neutral mid-layer of advertising, at the same time reinforcing the ideology of *caveat emptor*. If goods fail to satisfy, advertising culture asserts, the failure lies with the consumer. As Loeb (1994) has argued, this prescriptive format used images, for example school blackboards and accoutrements of the classroom, which rhetorically underscore this relationship. The luxury motif furthers this narrative. The association of particular goods with the abstract notion of luxury and with 'high' culture suggests that such goods are an 'acquired taste', appreciating them a sign of good taste, and thus conversely, finding them lacking a sign of *its* lack. Victorian adverts, then, created a complex set of subjectivities for their readers, in which the onus is always on the viewer to demonstrate their *own* credibility and prowess as a consumer, reversing the dynamic of disbelief.

In this respect, then, Victorian advertisements were a core element in the creation of the discourse of the consumer; in a Foucauldian sense, a structured set of subject positions, modes of behaviour, manners of activity and speech which all serve to bring forth the identity of the modern consumer (see Foucault, 1998). However, in another crucial respect, the advertisements articulated this identity, the consumer, with that of the citizen, and it is to this that we now turn in considering race and gender in Victorian advertising.

It is impossible to view nineteenth-century advertising with a modern eye and not be struck by the extent to which racialised images were invoked by advertisers. The range of products – from foodstuffs to beauty products to household goods – that were sold through the use of highly stereotyped images of race was broad, to say the least. It was, in fact, so prevalent that theorists have referred to what they call 'commodity racism', which is to say the conversion of imperialist narratives into consumer spectacle (McClintock, 1995). Soap advertisements have proved particularly fruitful as an object of study in illuminating the dynamics of this since soap as a commodity occupied a unique niche in Victorian cultural life. Ann McClintock, in her study of nineteenth-century promotions, has pointed to the fact that although soap advertising reiterated jingoistic and imperialist themes, soap itself was not linked to the empire project. Rather, it:

> emerged commercially during an era of impending crisis and social calamity, serving to preserve, through fetish ritual, the uncertain

> boundaries of class, gender and race identity in an social order felt to be threatened by the fetid effluvia of the slums, the belching smoke of industry, social agitation, economic upheaval, imperial competition and anticolonial resistance. Soap offered the promise of spiritual salvation and regeneration through commodity consumption.
>
> (1995: 211)

Soap adverts drew upon the lexicon of race to suggest and elaborate the idea of a link between the commodity and civilisation or social progress. This is particularly the case in adverts invoking the image of the white explorer or adventurer who was one of the key archetypes of nineteenth-century adverts (see above). The image of explorers like Henry Morton Stanley, author of the popular *In Darkest Africa* (1890), an account of his adventures and 'mission' to relieve Emin Pasha, was used to sell a large range of products. In one advert, the explorer was pictured finding comfort from his travails by drinking Bovril in the jungle; a little bit of home in the wild, the advertisement suggests (Richards, 1990: 143).

The theme of the contrast of 'home' and 'wild' is elaborated more directly in other advertisements. One particularly well-analysed advert for Pears soap pictures a tribal figure, standing on a beach, dressed only in a folded loincloth and carrying a spear. At the man's naked feet is a box embossed with the words 'Pears soap' from which several cakes of soap have fallen. The figure scrutinises one of the cakes whilst in the background a ship slowly sinks beneath the waves. The legend above the picture reads 'The Birth of Civilisation: A Message from the Sea', and beneath a message reads 'The consumption of soap is a measure of the wealth, civilisation, health, and purity of the people' (reproduced in Richards, 1990: 141). This advert enunciates several themes. In the first instance, it draws the obvious parallel between the commodity and (Western) civilisation. However, as Richards observes, the commodity is presented as a natural/neutral object since its 'message' requires no human agency to transmit. The sinking of the ship (and presumably its crew) in the background is the means by which the soap magically appears in the primitive world of the native. It is at a remove from precisely the dynamics of imperial conquest which gave it birth; the soap being a 'message' without a sender in this advertisement. At the same time, the aboriginal figure will, we are left to assume, work out the use of the soap and thus civilisation will be born.

Soap manufacturers could and did make heavy weather of dichotomies of white/black, clean/dirty in advertising their products. Another advert, again for Pears, featured a black child being bathed by a white child, the black child expressing surprise and delight when his black skin is 'washed white' below the neck, leaving his face as the only marker of ethnicity. However, they were not the only manufacturers to do so. Adverts consistently deployed the image of a racialised other in order to draw parallels with the supposed virtues of the white consumer. In this sense, as McClintock (1995) has argued, adverts invoked race as a means of policing class relations. The white consumer is less of a presence than an implied norm, and comedy is derived from the juxtaposition of the commodity and the ethnic subject. As McClintock has further argued, this is taken to an extreme in the common use of the image of the monkey in nineteenth-century advertising. The monkey is a liminal image belonging, as Haraway (1991) has argued, neither properly to the realms of the human nor directly to the animal world. As such, it is a powerful image as it literally transgresses boundaries, social, cultural and symbolic. The appearance of the monkey in advertisements of the nineteenth century acts as an exaggerated cipher of racial inferiority: it is a comic figure but one which is discursively imbued with quasi-human qualities and autonomy. In the same way, in the highly racialised discourse of Victorian advertising, the 'native' is figured as child-like, unintelligent, incapable of participating in the world created by the commodity. These images, it can clearly be seen, offered certain subject positions for the white audience, at whom the adverts were indeed directed. Failure to participate in consumption meant an automatic 'relegation' in racial identity, and questioned the status of the non-consumer as a citizen.

However, it is gender that forms the chief axis in advertisers' creation of the ideal consumer. As Loeb (1994) has noted, images of masculinity in nineteenth-century advertisements were attenuated and one-dimensional in comparison with the rich and detailed evocation of the female subject. In large measure, of course, this is a corollary of a domestic division of labour in which buying goods falls to women. However, it is also tied to the commercial appropriation of an elaborate domestic ideology that sanctified the home as a space of retreat from the public world and positioned women as the guardians of this 'haven of peace' through their roles as 'angels of the home'. As many historians have noted (Davidoff and Hall, 1987; Lewis, 1984), women's positions within the household were materially as well as culturally significant since a family's fortunes could rise and fall as a consequence of women's success in managing complex social performances in addition to arduous domestic tasks. For middle-class families, 'keeping up appearances' was vital in order to maintain their position and avoid a creeping proletarianisation. For poorer families, enacting displays of 'respectability' was one way to ensure their welfare in a labour market inimical to security (Ross, 1985). Advertisers drew on this rich seam to simultaneously flatter and unsettle their female viewers. Women were subject to a bombardment of images of the ideal woman as 'sweetness and light', in Matthew Arnold's terms (see Arnold, 1869/1960); all wise and knowing, whilst of course immaculately dressed. At the same time, advertisers continually reiterated themes tied to social anxieties, especially around fears of disease, ageing, child injury and illness, adulteration of foodstuffs and, above all, failure, leaving the consumer feeling 'unsure and ill-equipped' (Loeb, 1994, see above).

However, it would be wrong to see all Victorian advertisements aimed at women as only 'negative' in this sense. Victorian advertisers offered imagery of women that was in some respects liberating. Although an appeal to the image of the woman as a household angel offers only one arena in which women are pictured as experts, it was at least one. More importantly though, as noted above, in pursuit of a high cultural register, advertisements used imagery suggestive of classical themes, presenting women as commercial 'goddesses' by fashioning them in the style of statues of antiquity. Such themes drew upon imagery of privacy and retreat from the public space of course, but in place of the anxiety-driven matriarch, advertisers offered an image of sensuousness and robust physicality. The soft flowing lines and bared muscular arms of these figures were a telling substitution of the more usual depiction of the female body, buckled into restraining corsets and/or sheathed in layers of heavy cloth. There is a vision here not only of luxury but also of relief and easement. Such advertisements collude with the viewer, suggesting a knowledge of and sympathy with their struggles, offering a chance to 'unburden' themselves. At the same time, they also offer an image of feminine sensuality and private pleasure in embodiment. Advertisers then offered women a complex vision of themselves as consuming subjects, sometimes flattering, sometimes consolatory, more often tied, as noted above, to gendered anxieties and/or accompanied by a hectoring didactic tone which invited the reader to the position of 'student' of shopping.

Spaces of exhibition

The semiotics of advertisements is, of course, only part of the story. As the case of insurance has made clear, the dividing line in the Victorian era between the physical and the symbolic is blurred at best. In the same way that advertisements made the physical into symbolic, through the use of the imagery of the office and the building to solidify the product, so nineteenth-century adverts also blurred the lines in the other direction, making the physical environment into a symbolic one. In the first instance, it can be argued that the spaces of the exhibition of the commodity were intensely 'theatrical', constructed through the metaphor of the stage. The new shopping spaces that had emerged in large cities tended to be lavishly designed as retail palaces to 'stage' the new commodities. Theatre architect Frank Matcham, who is better remembered for his designs for the London Palladium and London Coliseum amongst his many other theatre designs, for example, designed the Victoria Quarter in Leeds. This is, of course, not a coincidence. As both Rappaport (2000) and Richards (1990) have pointed out, the spaces of consumption of commodities in the nineteenth century were as a much a part of their promotion as textual and pictorial advertisements, and this was never more apparent than in the case of the 1851 Great Exhibition. The exhibition was of course in many respects a propaganda exercise. It brought together manufactured goods of thirty-two nations under the roof of the purpose-built Crystal Palace in Hyde Park, London. Visitors could walk through exhibitions showing amassed manufactured products, divided into galleries based either on nation of origin, or, in the larger 'home' section, on type and use. The Exhibition was billed as the triumph and embodiment of a new liberal consumer era, one which demonstrated and embodied the equation of progress and material goods. A *Times* editorial of 1 May 1851 sums up the hopes and expectations of the Great Exhibition well:

> Till this day it has never occurred that all the nations of the earth should meet ...and combine in a common act. Divided by seas, by their still wilder passions, by language and by religion, the many races of mankind have been scattered over the surface of this globe, as over an illimitable waste...Other governments are ever mustering their legions, and overawing their neighbours and their subjects with the glittering pageant of war...(by contrast)...everything here is a pledge of peace, and shrinks from the rude grasp of war. Why should we scruple to congratulate the world that it has...(met here)...to exhibit the fruits of that industry, of that skill, of that earth which GOD has given to us for our common use and mutual obligation?

It was, however, 'England's good fortune to be the seat of this conquest' and her 'enthronement' at the head of the enterprise was, it argued, accepted by all nations who sought to come together in this 'Temple of Peace'. For the *Times*, as for many contemporaries, the exhibition of commodities was a statement of benevolent British dominance disguised as inclusivity. The exhibition of commodities and the hosting of such an event underscored British dominance, whilst restating the central liberal principles of *laissez faire* and market capitalism. The space of commodity exhibition, then, was a statement as much political as economic. The exhibition set the stage for future advertising and for the development of commodity culture. Richards argues that it 'inaugurated a way of seeing things that marked indelibly the cultural and commercial life of Victorian England and fashioned a mythology of consumerism that has endured to this day' (1990: 18).

The spaces of display of the commodity in real life were mirrored and extended by their symbolic display in the form of posters. Posters drew upon and elaborated the dramaturgical metaphor by linking the product into the modes of exhibition of the theatrical performance. This is particularly the case with the pictorial poster and what has come to be called 'poster art'. Poster art as a style and its dominant modes of representation came to nineteenth-century Britain via France, where posters had developed a wider currency and greater acceptance (Collins, 1985). The work of artists like Jules Cheret and later Henri de Toulouse-Lautrec, Théophile Steinlen, Eugène Grasset and Alphonse Mucha developed a new aesthetic in which formal conventions of art, especially perspective and composition, were disregarded in favour of movement and colour. Britain soon followed on from this in the work of many artists, most notably Frederick Walker in 1871. Walker's poster for the theatrical dramatization of Wilkie Collins' *The Woman in White* is generally accepted as the first British 'poster art'. Poster art is intended to arrest the eye and uses a simplified lexicon of images in order to grasp attention and convey meaning. As such, as Karen Carter (2012) has argued, they were consumed in a markedly different manner to other forms of public communication. Carter argues that advertising posters mark a transformation in the nature of spectatorship from the close group reading associated with civic notices to an 'internalised apperception' in which the spectator sees only in order to recognise rather than to understand. As noted above, the nineteenth-century metropolis was dominated by a wild proliferation of promotional texts. The mode of reading associated with this wall of images, Carter argues, was inattentive, paralleling a more generalised 'disengagement from public life and a larger atomization of the individual' (2012: 12). Poster art, then, colonises public space, individualising and depoliticising urban spaces into a theatre of images and objects, rather than, for example, spaces of assembly and debate. Adverts enshrined an aesthetic of looking without seeing.

Conclusion

Victorian advertising, in its many manifestations, was profoundly imbued with relations of power, being as it was constructed from markers of class and gender. As we have seen, the tropes and themes of Victorian advertising drew upon a sceptical and conservative vocabulary in which the boundaries of 'them' and 'us' were continually redefined and re-described. Thus, for example, the confessional form and the use of the tropes of the 'overheard' obfuscated the advertisement as a cultural artefact, redefining it as a service and interpellating the reader as the one 'in the know'. At the same time, Victorian advertisers drew on a longing for security to position themselves as intermediaries, negotiating the fraught and confusing world of commerce on behalf of the consumer. The invocation of the cosiness of the village green and the intimacy of friendship were put into service to transfer these affective bonds onto the advertiser. Advertisements sought to speak the language of the new consuming class, whether this was the desire to emulate the luxury of the fading aristocracy, or to grasp the respectability offered by a matriarchal role. At the same time, they were not afraid to appeal to jingoistic sentiments. Adverts reiterated racial stereotypes that offered the reader, the white consumer, a vicarious sense of citizenship and entitlement. Consumption marked the boundaries of citizenship of the 'empire', Victorian advertisements suggested. Finally, nineteenth-century advertisements reinforced existing relations of power through the colonisation of space. On the one hand, the spaces and places of advertising – from the 'poor man's galleries' of the streets to the transformations of city spaces into spectacles of display and consumption in the new shopping areas – reiterated the notion of advertising and the commodity as a demotic experience, in which all could participate through the activity of viewing. The styles of advertisements, including the reproduction of 'great art' and the new style of poster art, underscore this aesthetic of accessibility. At the same time, advertising posters added to a 'move along' culture in city spaces, appropriating and depoliticising the common places of the urban environment.

References

Althusser, Louis (1970) Ideology and Ideological State Apparatuses in John Storey (2009) *Cultural Theory and Popular Culture: A Reader*, 4th edition London: Routledge, pp. 302–12

Arnold, Matthew (1869/1960) *Culture and Anarchy* Cambridge: Cambridge University Press

Berry, Christopher (1994) *The Idea of Luxury: A Conceptual and Historical Investigation* Cambridge: Cambridge University Press

Carter, Karen (2012) The Spectatorship of the Affiche Illustre and the Modern City of Paris *Journal of Design History* 25 (1) 11–31

Collins, Bradford (1985) The Poster as Art: Jules Cheret and the Struggle for the Equality of the Arts in Late-Nineteenth century France *Design Issues* 2 (1) 41–50

Davidoff, Leonore and Hall, Catherine (1987) *Family Fortunes: Men and Women of the English Middle Class* London: Routledge

Foucault, Michel (1998) *The History of Sexuality, Vol. 1: The Will to Knowledge* Harmondsworth: Penguin

Haraway, Donna (1991) *Simians, Cyborgs and Women: The Reinvention of Nature* London and New York: Routledge

Hindley, Diana and Hindley, Geoffrey (1972) *Advertising in Victorian England 1837–1901* London: Wayland

Joyce, Patrick (1991) *Visions of the People: Industrial England and the Question of Class, 1840–1914* Cambridge: Cambridge University Press

Leiss, William, Stephen Kline and Sut Jhally (1997) *Social Communication in Advertising: Persons, Products and Images of Well-Being* 2nd Edition London: Routledge

Lewis, Jane (1984) *Women in England, 1870–1950: Sexual Divisions and Social Change* Brighton: Wheatsheaf

Loeb, Lori Anne (1994) *Consuming Angels: Advertising and Victorian Women* Oxford and New York: Oxford University Press

Macaulay, Thomas Babington (1830) 'Mr Robert Mongomery' in Terry R. Nevett (1982) *Advertising in Britain: A History* London: Heinneman

McClintock, Anne (1995) *Imperial Leather: Race, Gender and Sexuality in the Colonial Contest* London: Routledge

McFall, Liz and Dodsworth, Francis (2009) Fabricating the Market: The Promotion of Life Assurance in the Long Nineteenth century *Journal of Historical Sociology* 22 (1) 30–54

Marchand, Roland (1985) *Advertising the American Dream: Making Way for Modernity 1920–1940* London, Berkeley and Los Angeles, CA: University of California Press

Nevett, Terry R (1982) *Advertising in Britain: A History* London: Heinemann

Rappaport, Erika Diane (2001) *Shopping for Pleasure: Women in the Making of London's West End* Princeton, NJ and Oxford: Princeton University Press
Richards, Thomas (1990) *The Commodity Culture of Victorian England Advertising and Spectacle 1851–1914* London and New York: Verso
Ross, Ellen (1985). 'Not the Sort that Would Sit on a Doorstep': Respectability in Pre-World War I London Neighbourhoods *International Labour and Working Class History* (27, Spring) 39–59
Schudson, Michael (1984) *Advertising, the Uneasy Persuasion: Its Dubious Impact on American Society* New York: Basic Books
Smith, Woodruff D (2002) *Consumption and the Making of Respectability 1600–1800* New York and London: Routledge
Storey, John (2009) *Cultural Theory and Popular Culture: A Reader*, 4th edition London: Routledge
The Times (1851) Editorial *The Times* 1 May Issue 20790 p. 4
Twitchell, James (1996) *Adcult, USA* New York: Columbia University Press

7 Liminal seaside?

Working-class tourism in the nineteenth century

Robert Troschitz

Popular novels like Charles Dickens's *Oliver Twist* or Friedrich Engels's famous description of Manchester in *The Condition of the English Working Class* have created a rather bleak image of the nineteenth century. However, child labour, workhouses and dreadful living conditions are only one part of the story. Not only did the nineteenth century see increasing political democratisation and ground-breaking technological developments, but, with rising incomes and cheaper means of transportation, travelling for pleasure became possible for the many. While going on holiday had been the privilege of the few for centuries, a growing number of clerks, shop assistants and workers were now given the means to spend a few days of holiday away from home – and the place most of them went to was the seaside. Within only a few decades, the number of holidaymakers that populated Britain's shores grew tremendously – a development which can nowhere better be seen than in Blackpool, where the number of railway passengers increased from about 135,000 in 1861 to around 3 million in 1904 (Walton, 1983: 23). From the late nineteenth century onwards, the seaside holiday thus became an integral part of British popular culture.

Before analysing the popularisation of the seaside holiday in detail, one should pause for a moment and consider why the seaside holiday is such an interesting object of study in the first place. Without doubt, the holiday belongs to the world of leisure. But while other leisure activities like going to the pub, playing football or dancing in the music hall may be done on a weekly basis, going on holiday is something much more special. Instead of being a part of the everyday, the essence of the holiday is to break with the routines of everyday life (Berghoff and Korte, 2002: 3; Urry and Larson, 2011: 3). The holiday is the time-out in the course of the year. It is, as Fred Inglis calls it, "bracketed time" (2000: 9). But the holiday is not only a break with the everyday on a temporal level. Going on holiday also means to move out of our everyday surroundings and go someplace else – not any place, of course, but places that we desire. Tourist destinations are more than mere geographical locations. They are places with which we anticipate pleasure and that have particular meanings for us. Precisely in this respect, the seaside is also a rather special place that is distinct from other tourist destinations.

The seaside is not any kind of land. It is a borderland between land and sea, between the inhabitable and the uninhabitable, between civilisation and nature. According to Thomassen, "seaside and beaches are archetypical liminal landscapes" (2012: 21). But as catchy as this description of the seaside as a liminal landscape sounds, it raises the question what the term 'liminal' actually means.

The adjective 'liminal' and the corresponding noun 'liminality' derive from the Latin word *limen*, which means threshold. The concept of liminality was first introduced by Arnold van Gennep in his work on rites of passage and later taken up and popularised by Victor Turner. In this context, liminality refers to the middle stage in a rite of passage (Thomassen, 2012: 21). In this liminal stage, the subject is already detached from his or her earlier position and identity but is not yet reassigned a new one. As Turner later writes: "Liminal entities are neither here nor there; they are betwixt and between the positions assigned and arrayed by law, custom, convention and ceremonial" (1977: 95). Liminality thus refers to moments or periods of instability and discontinuity, the destruction of identity and the disruption of social order (Shields, 1991: 83). While Turner's work on rituals highlights the temporal dimension of liminality, it also has a spatial dimension. Thus, van Gennep also discusses 'the territorial passage' through liminal zones such as doorways or border areas between nations (Thomassen, 2014: 91). Such liminal spaces are 'places in-between' that allow for the suspension of social structures. As Preston-Whyte argues, "liminal spaces are intangible, elusive, and obscure. They lie in a limbo-like space often beyond normal social and cultural constraints" (2004: 350). Now the question arises to what extent the seaside is indeed such a liminal space where the social norms of everyday life are transgressed. If we have a look at descriptions of the seaside in the nineteenth century, we do find instances which clearly allude to the transgressive potential of the seaside. In *Porter's Guide to Blackpool* of 1871, we can, for instance, read the following:

> From the promenade to the water's edge the beach is gay with company, and the young and old, the healthy and infirm, the rich and the poor, are seen riding, walking, sitting, or reclining, in every variety of costume, along the sands or on the parade above.
>
> (1871: 45)

This is a remarkable image for its time. The seaside appears as a place that can be enjoyed by anyone, irrespective of age, health and, most importantly, wealth. In the society of the nineteenth century – a society that was deeply riven by class – the seaside seems to be the place where Benjamin Disraeli's 'two nations' – 'the rich and the poor' – are happily united as what Victor Turner called a *communitas* – a form of community of equals in which social structures and classifications are dissolved (1977: 96, 177). However, the very openness of the seaside that *Porter's Guide to Blackpool* praises can also be a source of conflict as different social groups with rather different intentions and understandings of the seaside holiday have to face each other and compete for recreational space. This is exactly what happened in the nineteenth century, and in particular the 1870s to 1890s, when the seaside holiday became a popular phenomenon and the working classes 'invaded' a social space that had so far been the reserve of the upper and middle classes.

After a short outline of the historical development of the seaside holiday, this chapter will investigate these conflicts and probe to what extent the seaside functioned as a liminal space that allowed for the transgression of everyday social norms and boundaries, in particular in terms of class, or whether the social confines that structured everyday life in Victorian Britain were actually transferred to the seaside.

The popularisation of the seaside holiday

Although sea bathing had been a popular custom for a long time (Walvin, 1978: 17–18; Williams, 2003: 27), the development of the seaside into a tourist destination and the construction of seaside resorts only began in the eighteenth century. At that time, the aristocracy and gentry increasingly discovered the seaside as an alternative to inland spas like Bath. Accordingly, the seaside was primarily visited for health reasons. This focus on health was promoted through treatises like Richard Russell's *A Dissertation on the Use of Sea Water*, which advised readers on how to use sea water to cure various diseases of the glands and made recommendations like the following: "the Cure is to be finished by cold bathing in the Sea, drinking every Morning enough Sea Water to procure two or three Stools a Day, immediately upon coming out of the Sea" (1760: 94). Tourists would thus drink seawater and bathe in the sea which at this time did not mean swimming but rather dipping, with men bathing naked while women would take to the waters in a bathing costume (Love, 2013: 20; Travis, 1997: 12–13). The pursuit of health continued to be of importance well into the nineteenth century – albeit sea water ceased to be drunk and the consumption of air became more important (Beckerson and Walton, 2005: 55–6).

With the rise of Romanticism in the early nineteenth century, a new understanding of the sea emerged. While the sea had not been of prime interest to the picturesque tourist (Hassan, 2003: 21), the paintings of William Turner and the writings of William Wordsworth and Lord Byron, who spoke of the sea as "boundless, endless and sublime, the image of eternity – the throne of the invisible" (1816: 92), turned the sea into the central object of the romantic gaze, and watching the sea became an integral part of the holiday experience for the romantic tourist. At around the same time, the seaside resorts developed their distinctive architecture. While early seaside resorts like Scarborough, Brighton or Margate had resembled the typical spa architecture with assembly rooms, theatres, coffee houses and libraries (Walton, 1983: 7, 157–8; Löfgren, 1999: 113), the construction of promenades and piers reinforced the status of the seaside as a non-ordinary place – a status which was further emphasised through buildings like Brighton's Royal Pavilion and West Pier, which transformed the seaside into an oriental fantasy (Gray, 2006: 92–4).

The greatest transformation that the seaside witnessed was, however, the arrival of the railway in the mid-nineteenth century. While travel by coach had been troublesome and expensive, the railway provided much faster and cheaper access to the sea. As a consequence, the number of seaside tourists rose with unprecedented speed. Brighton, for instance, saw a massive increase in tourist numbers after a railway connection to London had been opened in 1841. Within only thirteen years, traveller numbers to Brighton rose from 50,000 for the whole year of 1837 to 73,000 within a single week in 1850 (Walton, 1983: 22). The rising number of seaside tourists went hand in hand with a widening of the social spectrum. As wages were rising, the salaried middle classes could increasingly afford to spend one or two weeks at the seaside by the mid-nineteenth century (25). Only a few decades later, increased spending power also allowed parts of the working classes to travel to the shores (Walton, 1981: 249). However, the spread of the railway and rising incomes alone cannot explain the popularisation of the seaside holiday. A further precondition was the availability of free time. This actually explains why working-class tourism first appeared on a mass scale in Lancashire where traditional customs such as wakes or Whitsuntide had survived and provided several free days in a row (Walton, 1981: 254; Barton, 2005: 74–5). At times, these holidays were even supported by employers for whom uncontrolled mass-absenteeism was more of a problem than closing a mill for an agreed period of time (Walton, 1981: 255). The gradual rise of mass seaside tourism was thus the result of folk traditions, pressures from below as well as management methods (Berghoff, 2002: 163).

The growth of mass seaside tourism and the popularisation of the seaside holiday amongst the working classes were not without problems. Not only did the working classes enter a social space that had been dominated by the middle and upper classes, but, instead of adapting to the prevailing holiday practices, they also brought their own culture with them. As a result, conflicts emerged, and the seaside moved to the forefront of Victorian debates over the use of leisure.

The seaside and the use of leisure

As has been shown above, the seaside holiday as it developed in the eighteenth and early nineteenth century had primarily been about the pursuit of health and romantic views. This image was also endorsed in numerous guidebooks like Button's *The Brighton and Lewes Guide* (1805) or Theakston's *Guide to Scarborough* (1841). Likewise, *Porter's Guide to Blackpool* (1871) portrayed the seaside as a place "to restore vigour to the tired body" (2) and the sea as an "object of sublime grandeur" (3). In contrast to this official and promotional discourse of the guidebooks, the 'trip to Blackpool' stories of the 1880s paint a totally different picture of the seaside holiday. In John Hartley's *Seets I' Blackpool* (c.1883), for example, we learn more about popular amusement places like Uncle Tom's Cabin and Raikes Hall than about what the seaside has to offer in terms of health or romanticism (58–66, 82–5). In Benjamin Brierley's *Drop't on at Blackpool* (1881), Sam Smithies, Jack-o'-Flunters and Joe-o'-Nats sing a song in anticipation of their trip to Blackpool that mainly depicts the seaside as a place of fun. While the first line of the song actually refers to the health qualities of the seaside, the rest of the song neither focuses on health nor romantic views. Instead, Blackpool is all about "jackasses ridin' on donkey backs", "swings for young women to show their new boots", "drinkin' an' dancin'" and the "crackers and squibs" in Raikes Hall (Brierley, 1881: 5). Later in the same book, Ab-o'th'-Yate, who initially came to Blackpool for health reasons, concedes:

> We just did what lots besides us dun when they go'n to Blackpool for their health. We went to Uncle Tom's Cabin, to th' Strawberry Gardens, Cherrytree, an' places I've forgotten th' name on. We paraded th' piers till everybody knew us; an' then we finished up at Raikes Hall, wheere hunderts seemed as if they'd made up their minds just to have one fling i' their lifetime, an' then sattle deawn to steady wark an' sober livin'. (18)

These images stand in stark contrast to the ideal of the respectable and therapeutic seaside holiday. Blackpool is not presented as a place of leisurely strolls to restore the body, but as the home of German bands, fairs, theatres and places for drinking and dancing. This focus on fun and entertainment, which came with the working classes but was also growing amongst the middle classes (Croll, 2004: 404; Walton, 1983: 41), soon became a cause for conflict and triggered debates about the appropriate use of the seaside holiday.

One particular way of dealing with the rise of working-class tourism and the growing emphasis on entertainment can be found in *The Popular Guide to Blackpool* of 1889. The guide actually approves of the democratisation of the seaside holiday and also takes issue with the change of holiday practices. After pointing out the "glorious sea view" and the pleasures of bathing, the guide tells its readers the following: "But, over and above its natural beauties, and artificial aids to beauty, Blackpool takes high rank amongst places which cater for the amusement and entertainment of their visitors" (*Popular Guide*, 1889: 4). The guide thus acknowledges the rising demand for entertainment. Shortly after, we are, however, informed that amusement is subordinate to the pursuit of health, which is considered "paramount to pleasure, for without the one the other is impossible" (6). Moreover, amusements ought to be "pure and intelligent" as well as "elevating and refining" and should "sweeten and refine" one's life (4). Amusement is thus only appreciated if it improves and cultivates the tourist. In this respect, *The Popular Guide to Blackpool* continues the ideals of the rational recreation movement, which had sought to cultivate the working classes by transforming their leisure practices, for instance, through the provision of libraries and museums, penny readings and the organisation of sport (Beaven, 2005: 16–43; Bull, Hoose and Weed, 2003: 10–12). These ideals had also informed the early seaside excursions organised by Sunday schools, temperance societies and mechanics' institutes in the early Victorian years (Williams, 2003: 39; Walton, 1981: 249). The seaside holiday had been seen as a means to move the working classes away from the fairs of the industrial towns and to promote leisure practices that were educational, morally uplifting and enhanced health and fitness. But while the rational recreation movement had aimed for the improvement of the working classes, *The Popular Guide to Blackpool* also promotes the seaside holiday because "those who spend their leisure upon Blackpool's shores may put themselves within reach of influences which [...] make the toil and moil of work-a-day existence more easily to be performed with patience" (1889: 4). By allowing a break from the everyday life of work, the seaside holiday becomes a means to ensure the smooth functioning of industrial society.

The cultivation of the rising demand for entertainment was only one way of dealing with the growing popularisation of the seaside. Much more often, the working classes were treated with contempt. Thus, a resident of Southend complained in 1882 about the "the deplorable, drunken, and disgusting rabble that daily pours into Southend" (quoted in Churchill, 2014: 45). One of the reasons why locals might see the working-class tourists in such a negative light was their rather low economic significance. In particular, day-trippers did not bring any money to the resorts because they did not stay overnight and often took their own food with them (Churchill, 2014: 46). This is also depicted in a *Punch* cartoon of 1888, in which a head waiter resentfully remarks that the trippers are of no use to the trade as "they all bring their nosebags with 'em!" (Hammerton, c.1907: 167). Such a negative attitude towards the working-class tourist was not confined to locals but could also be found amongst other tourists. For example, a visitor to Blackpool condescendingly remarked in 1897: "They expect to see the attractions of their own towns (on market days and at Wakes) glorified, and that is what the sands have been" (quoted in Walton, 1983: 182). *Punch* also repeatedly took issue with the clash between different social groups and the transformation of the seaside into a noisy place that hardly allows for relaxation (Hammerton, c.1907: 62–8, 173). A cartoon of 1890, for instance, shows a crowd that goes mad on Brighton's promenade and thus makes it impossible to relax and restore one's health (171). The popularisation of the seaside and the growing number of 'nigger' minstrels, Punch and Judy shows, German bands and other popular forms of entertainment are seen as an intrusion and obliteration of the proper seaside holiday.

While this condemnation of the rise of working-class tourism may appear as rather different to the more favourable attitude towards the democratisation of travel in *The Popular Guide to Blackpool*, there is actually much common ground. No matter whether the aim was to improve working-class leisure practices or whether these were despised, both cases display the middle classes' unrivalled belief in their moral superiority – a perspective through which working-class leisure practices do not just appear as different but as inferior. In this respect, the popularisation of the seaside holiday was marked by the same class divisions that characterised everyday life in Victorian Britain, and the seaside did not function as a liminal space where class conflicts could be overcome. Indeed, class divisions were even further strengthened by the spatial organisation of the seaside.

Keeping the masses at bay

Although the seaside was open for everyone who could afford to travel, this did not mean that it became a shared space where different social classes would mix. As the working classes travelled at fixed times, for instance during wakes, it was easy for other tourists to avoid them (Barton, 2005: 76). Moreover, different resorts attracted rather different tourist and a resort hierarchy based on social tone developed. Thus, the middle and upper classes started to move to more remote places, which were out of the reach of the working-class tourists as these neither had the time nor the means to travel far (133). But remoteness alone cannot explain the development of a resort hierarchy. The efforts made by local government but also particular patterns of landownership also determined how resorts would develop and what kind of entertainment they would offer. Thus, Southport, which consisted of rather large estates, saw a more planned resort development with hotels and fine gardens cater for more respectable tourists than its nearby neighbour Blackpool, which was a town of many small freeholders and provided fertile ground for pubs, stalls and fairgrounds (Perkin, 1981: 70–83; Walton, 1998: 58–60). But while all these factors may help explain how the resort hierarchy developed, it is actually much more crucial to analyse how central social tone and thus social segregation were to the holiday experience.

How important social tone was to the Victorian seaside tourist is aptly depicted in a *Punch* cartoon which features Snobson, who has come to an "out-of-way seaside resort" and enquires about the "sort of people [...] you get down here in the summer" (Hammerton, c.1907: 81). Snobson actually appears to be afraid of mixing with people below his social status. Unfortunately, he soon has to learn that "there be fine people an' common people, an' some just half-an'-half, like yourself, zur" (ibid). Such a concern with social tone was not entirely new. We find it, for instance, in Charles Dickens's short story 'The Tuggses at Ramsgate', first published in 1836. When the Tuggs family discusses which seaside resort to go to, they dismiss Gravesend as "low" and Margate as "worse and worse" because there is "nobody there, but tradespeople" (Dickens 1996: 372). Social tone was thus already an issue when the middle classes first came to the seaside, but it became increasingly important with the rising number of working-class tourists.

While it may not be surprising that those who felt themselves socially superior were worried about mixing with the wrong people, a *Punch* cartoon of 1883 suggests that similar concerns could be found lower down the social scale as well. Thus, we find Chatty Visitor – a working-class chap who has come to Margate – explain to a rather astonished gentleman: "I like the place. I always come here. Worst of it is, it's a little too dressy!" (Hammerton, c.1907: 115). In other words, Chatty Visitor would feel much more comfortable if he was only surrounded by people of his own social class. As this cartoon indicates, the desire to stick with one's kind was a phenomenon that could be found amongst all social classes and not only those further up the social scale.

Besides the development of a resort hierarchy, some resorts like Blackpool also developed segregated zones for different types of tourists. While the beach as a public space was open to all and sundry, access to private spaces like gardens or piers could rather easily be restricted. Thus, *The Popular Guide to Blackpool* reassures its readers that "the Claremont Estate [is] the property of a private company" and can thus be "kept free 'from the madding crowd'" (1889: 10). Less affluent tourists could also be deterred by rather high entrance fees. In Blackpool, for example, the North Pier (opened in 1863) charged two pence and thus twice as much as South Jetty (opened in 1868). Moreover, South Jetty soon offered open-air dancing to a German band and thus attracted day-trippers and excursionists, leaving the North Pier to the better-off tourist (Walton, 1983: 39, 1998: 38–9; Cross and Walton, 2005: 24). While there may have been some places such as Blackpool Tower (opened in 1894) where different classes mixed (Walton, 1998: 92; Webb, 2005: 134), resorts like Blackpool were otherwise characterised by rather strong class divisions.

The crucial question is, however, not whether such forms of social segregation existed, but rather whether the tourists were conscious of them. As the 'trip to Blackpool' stories of the time suggest, there was indeed a deep awareness of these forms of social zoning. In *Drop't on at Blackpool*, for example, Ab-o'th'-Yate compares Blackpool's two piers:

> One finds a different sort o' folk upo' th' new pier [South Jetty] to what we meeten with upo' th' owd un [North Pier]. We dar goo closer to 'em witheaut fear o' bein' towd by looks, if not by words, to stond eaut o'th' road. Th' bacca [tobacco] reech smells strunger, an' th' women are leauder abeaut th' meauth. We know what class their return tickets are, an' could very nee calkilate how mony bottles they'd flung through th' carriage windows on th' road comin'. [...]
>
> Goo upo' th' owd pier, an' yo'n find things quite different. Everybody theere seems to be sayin', 'Keep off my toes;' an' there's no creepin' like snails fro' one end to th' tother, an' laafin' at nowt. Theere noses are

> cocked up, an' legs are swingin' in a hurried walk, as if they'rn expectin' a sheawer – for sometimes it rains i' Blackpool – or th' pier givin' way afore they geet back.
>
> (Brierley, 1881: 14)

In *Seets I' Blackpool*, Sammywell Grimes, though not comparing the two piers, also reflects on the different clientele to be found in different places. On the North Pier, he encounters:

> old fowk at lukt as if they nivver had a care, an' young ens 'at lukt too beautiful an' pure to be classed amang 'miserable sinners,' an' rosy little rollickin' lads an' lasses, soa full o' frolic, soa free o' care, at one wished it wor possible to keep em just as they wor.
>
> (Hartley, c.1883: 34)

At Uncle Tom's Cabin, the scenery is the complete opposite:

> Ther wor a famous lot o' fowk abaat, net quite sich a touch-me-not lukkin' set as ther wor abaat th' North Pier, an' they'd browt ther own baskets o' prog, an' wor suppin' ginger beer aght o' th' bottle neck, an' when they laft, they didn't hold up a fan ith' front o' ther face an' titter [...].
>
> (59)

Although one cannot read these fictional accounts as truthful representations of the tourist population, they suggest that the social zoning within Blackpool was strongly felt.

The establishment of a resort hierarchy and the social zoning within resorts actually prevented the seaside from becoming a liminal space where social boundaries would easily be overcome. Instead of being a space where "rich and poor" would mix, as *Porter's Guide to Blackpool* had suggested, the class distinctions that characterised everyday life in Victorian Britain were maintained at the seaside. The great concern with class and social distinction actually shows how central class divisions were to the Victorian society and how impossible and dangerous it appeared to allow for a place where these divisions would dissolve. However, class boundaries were not the only area in which the seaside holiday perpetuated the social confines of everyday life.

The seaside holiday and the everyday

As has been argued above, the break with the everyday is the very essence of the holiday. However, in many respects such a break did not take place. First of all, the seaside holiday easily established a routine of itself. As Silloth's local newspaper aptly commented in 1860:

> We eat, we drink, we bathe, we walk, we sleep; and then we eat and drink and bathe and walk and sleep again. Only for variety's sake sometimes we bathe before we walk, and sometimes we walk before we bathe.
>
> (quoted in Walton, 1983: 169)

This monotony of the middle-class seaside holiday was also repeatedly satirised in *Punch* (Hammerton, c.1907: 10–16, 108, 132, 139, 142). While this routine is still based on practices particular to the seaside, it is remarkable that many everyday activities were continued at the seaside. Thus, a cartoon in *Punch* shows a number of men who are all reading the morning papers and thus not only keep in touch with their everyday surroundings but also repeat an everyday activity (19). The issue of reading is also taken up in *Seets I' Blackpool*, where Sammywell Grimes says to himself: "Ther's one thing aw connot understand abaat visitors to th' seah side. They all seem to be smittled wi a readin' fit, as if they couldn't read at hooam" (Hartley, c.1883: 57). This transferral of the everyday onto the seaside did not stop with activities such as reading. As we are informed in J.W. Kneeshaw's *A Black Shadow*:

> Men are as earnest and pushing here in search of pleasure as they are at home in pursuit of wealth. On all sides it can be observed. There is no rest, no dolce far niente. The day's work of pleasure-seeking is as carefully planned out beforehand, as if the day's bread is to be earned by it [...].
>
> (1894: 21)

All these descriptions suggest that the seaside holiday was no fundamental break with the everyday.

For the working classes, the continuation of the everyday into the holiday even went further. In a resort like Blackpool, the lodging houses often resembled the working-class houses of the industrial towns – a circumstance which John K. Walton sees as one of the reasons for Blackpool's popularity as the working classes could actually feel at home there (1998: 69). At the same time, the lodging houses imposed a temporal discipline that resembled factory life (Walton, 1998: 79; Webb, 2005: 127). But most importantly, working-class tourists travelled en masse with whole communities going to the seaside at once. Even if they were not taking part in an excursion, the working-class tourists would hardly be among strangers in a resort like Blackpool. This is again well depicted in the 'trip to Blackpool' stories. In *A Black Shadow*, Mary and Mrs Denton run into a number of familiar faces (Kneeshaw 1894: 23). In *Seets I' Blackpool*, Mally Grimes has to learn that being in Blackpool "wor ommost like bein' at hooam for ivverybody shoo met 'at shoo knew" (Hartley, c.1883: 46), and later Mally and Sammywell are told by a carriage driver that "if aw want to goa whear aw'st net be known, aw munnot come to Blackpool" (79).

While for some tourists this collective travelling may have been the essence of the seaside holiday as it kept the wakes-like atmosphere, Mally Grimes experiences it as a form of pressure. When her husband Sammywell suggests having a ride in a carriage, Mally answers:

> Tha's made up thi mind to goa aw know vary eel, soa what's use o' axin' me? Th' best thing aw can do is to get on mi' bonnet an' shawl an' goa wi' thi, but aw do hooap at ther'll nubdy see us, for if they'd to get to know i' Bradforth 'at thee an' me had been seen ridin' in a carriage they'd shaat us through th' streets.
>
> (Hartley, c.1883: 78)

As can clearly be seen, Mally is afraid of doing something out of the ordinary in case someone she knows might see her. Her home community exercises what Walton has called a "censorious collective gaze" (1998: 72), and thus prevents the seaside from becoming a liminal space where the social norms of the community could be transgressed. However, this does not mean that the seaside holiday did not allow for some deviant behaviour. Even anxious Mally finally dares to do something she would never do at home. When Sammywell invites her to go to the theatre, she replies: "Thee pleas thisen; but tha knows aw dooant hold wi' theayters nor places o' that sooart, an' if aw wor at Bradforth aw wodn't be seen gooin into sich a place, for awr praicher says they're associated wi' wickedness" (Hartley, c.1883: 114). Although Mally is again afraid of being seen, she finally goes into the theatre because "bein' away throo hooam aw dooant suppooas it matters whether we goa or net" (ibid).

Conclusion

The industrialisation of Britain had an enormous impact on the lives of the people and fundamentally reshaped working and living conditions. As part of this great transformation, leisure practices altered and new popular practices such as the seaside holiday emerged. While the seaside holiday had for long been the reserve of the few, the spread of the railway and increasing spending power brought the seaside within the reach of the middle and later also the working classes. No matter whether it was for a couple of weeks or just for a day, an ever increasing number of people now flocked to Britain's shores to leave the everyday behind.

However, this break with the everyday was only partial. The seaside holiday did not break with the social structures that shaped Victorian Britain, in particular in terms of class. Not only was the debate on the use of leisure marked by a strong class consciousness, but the resort hierarchy and the social zoning within the seaside resorts minimised class contact and prevented the seaside from becoming a liminal space where the class antagonisms of Victorian Britain would be dissolved. Indeed, the overarching concern with class shows how central class divisions were to the Victorian society and how impossible and dangerous it appeared to allow for a place where these boundaries would be overcome.

But class structures were not the only area in which the seaside holiday continued the social confines of everyday life. Besides the fact that the seaside holiday could easily establish an everyday routine of itself and that many everyday activities were not abandoned with at the seaside, the very system of working-class travel and the resulting collective gaze ensured that social norms were in place. Yet, this does not mean that the seaside holiday was the same as everyday life. The seaside holiday was, of course, a break with the life of work, and as we see in the 'trip to Blackpool' stories like *Seets I' Blackpool*, the seaside holiday was certainly a joyful and more excessive time than the everyday life back home. It is this special atmosphere of the seaside which explains its popularity and which makes Sammywell Grimes conclude: "if we are spared to live it's net likely "at this will be th' last trip to Blackpool after all" (Hartley, c.1883: 123–4).

References

Barton, Susan (2005). *Working-Class Organisations and Popular Tourism, 1840–1970*. Manchester: Manchester University Press.

Beaven, Brad (2005). *Leisure, Citizenship and Working-Class Men in Britain, 1850–1945*. Manchester: Manchester University Press.

Beckerson, John, and John K. Walton (2005). 'Selling Air: Marketing the Intangible at British Resorts.' *Histories of Tourism: Representation, Identity, and Conflict*. Ed. John K. Walton. Clevedon: Channel View. 55–68.

Berghoff, Hartmut (2002). 'From Privilege to Commodity? Modern Tourism and the Rise of the Consumer Society.' *The Making of Modern Tourism: The Cultural History of the British Experience, 1600–2000*. Eds. Barbara Korte, Christopher Harvie, Ralf Schneider and Hartmut Berghoff. Basingstoke, New York: Palgrave. 159–79.

Berghoff, Hartmut, and Barbara Korte (2002). 'Britain and the Making of Modern Tourism: An Interdisciplinary Approach.' *The Making of Modern Tourism: The Cultural History of the British Experience, 1600–2000*. Eds. Barbara Korte, Christopher Harvie, Ralf Schneider and Hartmut Berghoff. Basingstoke, New York: Palgrave. 1–20.

Brierley, Benjamin (1881). *Drop't on at Blackpool*. Manchester: Abel Heywood.

Bull, Chris, Jayne Hoose, and Mike Weed (2003). *An Introduction to Leisure Studies*. Harlow: Prentice Hall.

Button, J.V. (1805). *The Brighton and Lewes Guide: Containing an Abridged History of those Towns, with Descriptions of the Edifices, Amusements, &c.* Brighton: Baxter.

Byron, Lord (1816). *Childe Harold's Pilgrimage: Canto the Third*. London: John Murray.

Churchill, David (2014). 'Living in a Leisure Town: Residential Reactions to the Growth of Popular Tourism in Southend, 1870–1890.' *Urban History* 41(1): 42–61.

Croll, Andy (2004). 'Popular Leisure and Sport.' *A Companion to Nineteenth-Century Britain*. Ed. Chris Williams. Malden, MA: Blackwell. 396–411.

Cross, Gary S., and John K. Walton (2005). *The Playful Crowd: Pleasure Places in the Twentieth Century*. New York: Columbia University Press.

Dickens, Charles (1996). *Sketches by Boz*. London: Penguin.

Dickens, Charles (2003). *Oliver Twist*. London: Penguin.

Engels, Friedrich (2009). *The Condition of the English Working Class*. London: Penguin.

Gray, Fred (2006). *Designing the Seaside: Architecture, Society and Nature*. London: Reaktion.

Hammerton, J.A. (c.1907). *Mr. Punch at the Seaside*. London: Educational Book Company.

Hartley, John (c.1883). *Seets I' Blackpool, Fleetwood, Lytham and Southport: As Seen bi Sammywell Grimes an' his Wife Mally on their Hallidy Trip, wi' a few Incidents an' Accidents 'at Occured*. London: W. Nicholson & Sons.

Hassan, John (2003). *The Seaside, Health and Environment in England and Wales since 1800*. Aldershot: Ashgate.

Inglis, Fred (2000). *The Delicious History of the Holiday*. London: Routledge.

Kneeshaw, J.W. (1894). *A Black Shadow: A Tale of Lancashire Life*. London: Simpkin, Marshall, Hamilton, Kent & Co.

Löfgren, Orvar (1999). *On Holiday: A History of Vacationing*. Berkeley: University of California Press.

Love, Christopher (2013). *A Social History of Swimming in England, 1800–1918: Splashing in the Serpentine*. New York: Taylor and Francis.

Perkin, Harold (1981). *The Structured Crowd: Essays in English Social History*. Brighton: Harvester.

Porter's Guide to Blackpool, Fleetwood, Lytham &c. (1871) Blackpool: W. Porter.

Preston-Whyte, Robert (2004). 'The Beach as a Liminal Space.' *A Companion to Tourism*. Eds. Alan A. Lew, Colin M. Hall, and Allan M. Williams. Malden, MA: Blackwell. 349–59.

Russell, Richard (1760). *Dissertation on the Use of Sea Water in the Diseases of the Glands: Particularly the Scurvy, Jaundice, King's-evil, Leprosy, and the Glandular Consumption*. London: W. Owen.

Shields, Rob (1991). *Places on the Margin: Alternative Geographies of Modernity*. Abingdon: Routledge.

The Popular Guide to Blackpool (1889). London: Doherty & Company.

Theakston, Solomon W. (1841). *Theakston's Guide to Scarborough: Comprising a Brief Sketch of the Antiquities, Natural Productions, and Romantic Scenery of the Town and Neighbourhood*. Scarborough: Theakston.

Thomassen, Bjørn (2012). 'Revisiting Liminality: The Danger of Empty Spaces.' *Liminal Landscapes: Travel, Experience and Spaces In-Between*. Eds. Hazel Andrews and Les Roberts. Abingdon: Routledge. 21–35.

Thomassen, Bjørn (2014). *Liminality and the Modern: Living through the In-Between*. Farnham: Ashgate.

Travis, John (1997). 'Continuity and Change in English Sea-Bathing, 1730–1900: A Case of Swimming with the Tide.' *Recreation and the Sea*. Ed. Stephen Fisher. Exeter: University of Exeter Press. 8–35.

Turner, Victor (1977). *The Ritual Process: Structure and Anti-Structure*. Ithaca, NY: Cornell.

Urry, John, and Jonas Larsen (2011). *The Tourist Gaze 3.0*. Los Angeles, CA: SAGE.
Walton, John K. (1981). 'The Demand for Working-Class Seaside Holidays in Victorian England.' *The Economic History Review* 34(2): 249–65.
Walton, John K. (1983). *The English Seaside Resort: A Social History, 1750–1914*. Leicester: Leicester University Press.
Walton, John K. (1998). *Blackpool*. Edinburgh: Edinburgh University Press.
Walvin, James (1978). *Beside the Seaside: A Social History of the Popular Seaside Holiday*. London: Penguin.
Webb, Darren (2005). 'Bakhtin at the Seaside: Utopia, Modernity and the Carnivalesque.' *Theory, Culture & Society* 22(3): 121–38.
Williams, Stephen (2003). *Tourism and Recreation*. Harlow: Prentice Hall.

8 Shocking readers

The genres of Victorian popular fiction, the classes and the book markets

Ralf Schneider

Introduction: popular reading, reading popularity

In the days before radio, TV, cinema and the Internet, reading was one central activity with which people filled their leisure hours. The culture of everyday life in the nineteenth century comprised the reading of a broad spectrum of texts, ranging from religious edification to horror stories, and from conduct books to romance novels. Although people read broadly in all genres, prose fiction was dominant. The nineteenth century can in fact be thought of as the age of the novel. Students of the Victorian period are faced with the remarkable figure of some '60,000 works of adult and juvenile fiction published 1837–1901' (Sutherland 1988a: 1) – and that figure does not even account for the first three and a half decades of the century. As Sutherland goes on to say, 'there were, at an educated guess, around 7,000 Victorians who could legitimately title themselves novelists' (ibid.), and towards the end of the century at least 2,000 new novels for adults, children and adolescents were published every year; that is, more than five per day (Keating 1989: 32–33). Had there been no readership for that amount of fiction, it would not have been printed, so surely the nature of Victorian fiction generally was that it was *popular*?

Things are, however, more complicated, for the literary world of the Victorian period was subdivided – or subdivided itself – into numerous fields, ranks and niches, and the borders between them could be crossed in many ways as the century progressed. Who read what depended on the quality and quantity of education people had received, on the availability of texts, on leisure time and other things we now take for granted, such as light and a room to read in; and of course, reading depended on the budget a person had available and wished to spend for that particular recreational pursuit. All of this was connected with the class to which a reader belonged. This chapter is going to argue that the social stratification of British society in the nineteenth century on the one hand, and the literature of the period on the other, were intertwined in variegated and changing ways, and that popular literature was a highly complex phenomenon shaped by a set of aesthetic, ideological, economic and material factors that generally affected the book market, book production, audiences and authors in the nineteenth century (Sutherland 1976a, Feather 1988: 129–179, Keating 1989, Erickson 1996, McKitterick 2009). The story that this chapter is going to tell is that one variety of 'popular' literature was synonymous with literature written for the working classes and sold cheaply at the beginning of the century, and that, as the novel became both more attractive – and thus, 'popular' – to readers from all classes, middle-class readers of the previously more expensive fiction increasingly turned to formats originally intended for the low-price market. Over the Victorian period, changes in education, in society at large and in the book market led to a democratisation of literature, which created a situation that prevailed throughout the twentieth century: the novel in general became *the* popular literary genre, catering to the tastes of the broad population and attracting readers from all classes. At the same time, writers keen on advancing the aesthetic achievements of novel writing came to regard themselves as an avant-garde, followed by fewer readers in comparison to the bestsellers. Literary production bifurcated into a popular field (which included also what was later called 'middlebrow' literary production) on the one hand, and an elite field on the other.

It is safe to say that at the beginning of the century, there were at least two popular literatures, in the plural. A market for novels popular with the middle classes existed side by side with a fairly separate market of literature produced for the working classes – the latter being one meaning of the term 'popular' in cultural studies (Storey 2015: 5–14). Acknowledging that 'class' is just as problematic a term as 'popular' (Kidd & Nichols 1998), I follow the sociological notion of class divisions supported by strategies of 'distinction' (Bourdieu 1984) in this chapter. This approach takes account of the fact that practices such as reading habits, and conditions such as education, are relevant factors in the class-bound consumption of cultural artefacts. As subsequent sections will explain in more detail, the two markets were at first separated by subject matter and style on the one hand, and price levels of the literary products on the other (Suvin 1982): 'penny dreadfuls' and 'shilling shockers', for instance, were stories of gothic horror, crime, and the supernatural, cheap and mass-produced for the working classes. The middle classes in contrast were the target audience of the realist novel, mostly published in three volumes (called 'three-deckers' for their size, inspired by large ships with three decks), which were more decent in morality, much less lurid and more convincing in terms of plot construction and character development. Although they were too expensive to be bought even by members of the middle class, these novels were available from circulating libraries whose membership fees would deter the lower ranks. Initially, then, different types of fiction were popular with different classes, but there were also a number of developments indicative of a democratisation of the book market, which made the division between middle- and working-class markets increasingly permeable.

Cheap literature: penny dreadfuls, shilling shockers, chapbooks

Popular literature of the nineteenth century continued doing what much popular writing had done since the Early Modern period: it provided shocking stories of crime and mystery, vice and punishment (cf. Neuburg 1977). The eighteenth century had seen the introduction not only of a modern consumer society in which the acquisition and possession of goods became an end in itself, but also the introduction of laws that protected property by making property-connected crimes such as theft, shoplifting and poaching capital offences, to be punished with the death penalty (Thompson 2011 [1975]). While in the nineteenth century these laws were alleviated, the fascination with crime, criminals and (capital) punishment continued to thrive, and news journalism as well as prose fiction knew how to capitalise on people's hunger for horrid stories, whether real, exaggerated or entirely imaginary (cf. Anglo 1977). In the eighteenth century and before, the easiest and cheapest way for the less affluent readers all over Britain to get access to stories that narrated incidents of crime and punishment had been to buy broadsheet ballads or chapbooks from itinerant peddlers, or on markets and fairs. In these formats, usually sporting rather crude woodcut illustrations, people found stories which frequently included horrid details of exaggeratedly violent crimes as well as the gruesome deaths the delinquents met through the hands of executioners. When industrialisation brought a growing number of workers to the industrial cities with their exploding population figures, a new urban readership for these texts grew.

Different formats of such literature were within the reach of a worker's income, from the penny dreadfuls (also called 'penny bloods' before the 1860s) to the more expensive shilling shockers (one shilling was twelve pence), sold in chapbooks of various length. The cheapest booklets had eight pages, sometimes 16; longer chapbooks were slightly more expensive and had 36 pages for sixpence or 72 for a shilling. Penny dreadfuls came in instalments of eight pages length, which means that they could be read even by those whose reading skills would not induce them to extensive novel reading. The crime and gothic horror that was the central topic of the 'bloods' found many readers among the industrial workers: 'With its antecedents in the street ballad, the gothic novel and the *Newgate Calendar*, the penny blood enjoyed popularity among the urban working classes throughout the 1840s' (Powell 2004: 45). They were mostly produced as serials attracting readers over many months, and though the bulk of these works has been lost, they constituted a massive sector on the print market in the first half of the nineteenth century (see Anglo 1977 and the British Library homepage on penny dreadfuls [Flanders, n.d.] for access to interesting material, and cf. James 1963: 28–44 and Terry 1983). The most popular stories of this category include *Varney, the Vampire; or, The Feast of Blood* (1847–48) and the story of perhaps the most famous English murderer, eternalised in the series *Sweeney Todd, the Barber of Fleet Street* and so popular as to be frequently rewritten and dramatised (Neuburg 1977: 172). The *Varney* series, published with the caption 'A ROMANCE OF EXCITING INTEREST' on the front page (see illus. ibid.: 173), was written by James Malcolm Rymer for the young publisher Edward Lloyd. Lloyd made a fortune with penny publications and cheap newspapers that allowed him to buy himself into a more respected market by acquiring the *Daily Chronicle* in 1876. Coming from an uneducated background himself and knowing what people would want to read, he 'fully understood the market which he was successfully exploiting' (Feather 1988: 157). Rymer, like many of his fellow writers, was amazingly prolific, being the author of at least 120 works (James 2006: 138). It is obvious that an output of that scale was not likely to invite careful research, well-crafted plotting and meticulous editing, not least because writers who made their living from the writing of such stories would never finish one book before they began the next. While he was writing *Varney*, Rymer also worked on the long-running and extremely successful series *Mysteries of London*, which began 1844, continued over 12 years and amounted to '624 numbers and nearly 4.5 million words' (Flanders n.d.). The vampire story itself:

> ran to 220 chapters, nearly 900 pages, and it is not at all surprising that *Varney*, written over a long period by an author engaged in turning out episodes for other serials at the same time, lacks cohesion and is vague as to time and place.
>
> (Anglo 1977: 16)

It is easy to dismiss such style in terms of poor skills and substandard aesthetics. But what *was* the aesthetic standard against which such writing ought to be measured? Apparently, the publishers were well aware of the quality of their products, but it did not trouble them, for the target audience was apparently not believed to know – or even deserve – better. One of Lloyd's managers is reported to have said:

> Our publications circulate among a class so different in education and social position from the readers of the three-volume novels that we sometimes distrust our own judgment and place the manuscript in the hands of an illiterate person – a servant, or machine boy, for instance. If they pronounce favourably upon it, we think it will do.
>
> (quoted in Neuburg 1977: 172)

The manager is not only fully aware of the division of the book markets as a side effect of class divisions; to call a young person who can, after all, read the manuscript of a story 'illiterate' also betrays the level of condescension typical of middle-class producers of working-class literature. The standard was, basically, that as long as those with a penny to spend found the story exciting enough, delicacies of style were a dispensable luxury.

Other publications for the semi-literate working classes were slightly more expensive, but also concerned with horror and lurid crime, such as the shilling shocker and the chapbook (Watt 1967). The chapbook, also called bluebook in the nineteenth century and 'of a slightly higher order, selling for sixpence and a shilling' (10), frequently contained abbreviated and simplified versions of longer novels sold in other formats. The demand for such products was so high that printers and publishers even hired authors to pirate existing shilling shockers which could then be re-printed as sixpenny bluebooks. Pirating was easy at the time, because the copyright laws were far from satisfactory from the point of view of the authors (see Rose 1993).

They were about four by seven inches in size, and their closely printed pages were poorly stitched into a cover of flimsy blue paper. The stories varied in length from mere anecdotes to tales of 30,000 words, but many of the publishers specialised in two definite lengths, dealing out 36 pages for sixpence and 72 for a shilling (11).

The very physical appearance of such fiction came to signal what kind of book(let), qualitatively, one was buying. Beside readers' appetite for shocking stories, a number of interwoven factors brought cheaper texts to more and more people over the century. These included progress in printing technology and paper-making, particularly in the last three decades of the century, in combination with sinking costs for all materials and machines involved (Dooley 1992, Eliot 1994, Weedon 1996, Banham 2007); the removal of taxes on advertising, paper and newspapers (cf. Hewitt 2013); a rise in literacy (Mitch 1992, Eliot 2007, Vincent 1983 and 1989); the reduction of working hours; and rising wages for workers in the industry (Nelson 2005, Altick 1989, p. 148; Erickson 1996: 143).

Middle-class tastes and middle-class fiction: between serious and sensational

As we have seen, the fiction produced for the working classes secured sales by giving readers shocks of excitement of all kinds. While many producers of such stories were members of the middle classes, the working-class readers and their stories were considered shocking phenomena themselves by many middle-class observers. The leisure activities of the working classes had always been contested ground (Walton & Walvin 1983, Golby & Purdue 1999 [1984]), and reading was no exception. Middle-class fears of the dangers of working-class unruliness, alcoholism and indecency shaped the regulation of the workers' pastimes by the authorities. Other leisure activities competed with literature for the free time that workers had at their disposal, and the pub, animal fights and dog races were particularly suspicious to the guardians of morality; later in the century, music halls and penny cinemas added to the many other attractions, such as funfairs and pleasure gardens. There were many fears that mass reading, especially in the form of cheaply produced crime, horror and sensation stories led to moral decay and was 'rotting the minds of their readers, promoting vice, and subverting cultural standards' (Brantlinger 1998: 2; cf. also Altick 1973: 190–202, and Springhall 1994). Therefore, some movements and institutions, such as the religious evangelical movement, the Society for the Distribution of Useful Knowledge or the Mechanics' Institutes, endeavoured to diminish illiteracy among the workers, and supplied the working classes with morally uplifting or educational works, and with helpful guides for self-improvement. But it was not to be expected that all working people wanted to read for utilitarian purposes in the few leisure hours they had.

Many onlookers from the middle-class literary field reacted to working-class reading with alarm, others with disgust. George Eliot (Mary Anne Evans), one of the most eminent authors of serious middle-class novels, complained to her publisher, John Blackwood, in 1860 that 'the most carefully written books lie, both outside and inside people's minds, deep undermost in a heap of trash' (quoted in Maunder 2005: 1). This attitude is very much in line with a widespread evaluative take on practically everything that is available for 'the masses' which we find in many Victorian discourses and that dates back to the eighteenth century: literature that appeals to very many readers must be unworthy of critical appreciation. Middle-class observers were keen to keep their conception of literature distinct from that of the working classes, and middle-class literature developed means of distinguishing itself from that of 'the masses'. It is of course easy to be dismissive about the quality of cheap texts if one can afford expensive ones, and has learned to appreciate the world-views propagated in them. Some commentators were more benign in their attitude towards the working classes, but they still regarded them as a distinct set that remained potentially dangerous and in need of middle-class interference. Wilkie Collins, the author of extremely successful mystery and sensation novels of the 1860s (such as *The Woman in White* or *The Moonstone*), published an essay in 1859 on the readers of penny literature titled 'The Unknown Public' (Brantlinger 1998: 17). Much like William Makepeace Thackeray had done some 20 years earlier, Collins speaks of a 'monster audience of at least three millions', which confronts society with the task 'to teach that public how to read', because its members 'are evidently, in the mass, from no fault of theirs, still ignorant of almost everything which is generally known and understood among readers whom circumstances have placed, socially and intellectually, in the rank above them' (quoted in Brantlinger 1998: 217; cf. Keating 1989: 401–404). The middle-class men who habitually warned against the dangers of cheap literature for practically everyone except themselves betrayed some nervousness as to the likelihood that their wives and children might join the working class in their reading tastes: 'To the guardians of morality, a low price meant availability to women, children, and the lower orders, all of whom were much more corruptible than middle-class men' (Eliot 2007: 301f.). There was a barrier between middle- and working-class literatures, but if that had been a tight and secure one, no such nervousness would have been necessary.

Let us inspect how the barrier was erected and then turn to how it was overcome in the last section. As with the working-class reading, a close connection existed between book format, style, prices and modes of distribution in the literary market of the middle classes. Although some eighteenth-century authors had already tried out the publication of novels in three volumes, it was the success of Sir Walter Scott's long historical novels in three volumes, which were simply too long to be bound in one volume, that turned this format into a standard attractive to publishers (see Griest 1970, Sutherland 1976b, and Eliot 1985). This is connected to the circulating libraries, which had become the main platform for the distribution of new novels since their invention in the eighteenth century: novel readers were not novel buyers in the nineteenth century. The leader of the circulating library market was Charles Edward Mudie, with his Mudie's Select Library, as Griest (1970) shows in her comprehensive study (cf. also Erickson 1996: 125–141) on pre-Victorian circulating libraries. Mudie guaranteed publishers to stock a considerable number of the first print run of a novel in exchange for a reduced price. Indeed, his monopoly was so strong that he could afford to offer publishers only 15 shillings per novel, while the standard price was 31s/6d. This price made it impossible for all but the *very* affluent to actually buy novels. For instance, a London office clerk would have to invest a whole week's salary if he wanted to possess rather than borrow a novel (Altick 1989: 146–147). Even the well-to-do middle classes would usually not spend more than 6s, 7s/6d or a maximum of 10s/6d for a book (Griest 1970: 40). Contemporary observers estimated that only 60,000 of the 4.6 million British families could be considered 'comfortable'. Of those, the majority in fact subscribed to one circulating library or another: in the 1850s, some 50,000 persons were subscribers (Thompson 1996: 124), and some 40,000 families in the years 1890 to 1894 (Griest 1970: 78–79): obviously, there was a close correlation between economic position and novel-lending, which amounted to an exclusion of the working classes from that market segment. Mudie exerted much pressure on publishers to offer him only novels that met an almost puritan standard of decency, but his clients actually found a broad spectrum of books on the shelves of his circulating library, ranging from serious to sensational, and from instructive to trivial (135; Terry 1983: 10). The division between middle-class serious literature and working-class sensationalism was, after all, perhaps less a matter of fact than a matter of wishful thinking on the part of some middle-class supporters of aesthetic taste and social distinction, after all. There were three phases in which this division appeared less clear than some observers might have wished, as I am going to explain in the next section. Thackeray's and Collins' shock of discovering an entirely distinct group of readers between the 1830s and the 1850s is a bit surprising, because middle-class readers had also appreciated other fictions than serious three-volume novels, and Collins' own genre of the sensation novel caused a fierce debate about literary quality and moral standards (see below).

The democratisation of reading, or, popularising the popular

The process of democratisation that changed the political and social structures of nineteenth-century Britain was slow, halting and only reluctantly admitted by the upper and middle classes, but it eventually led to better working conditions, better representation of the people's interests in parliament, the extension of the right to vote and other reforms that acknowledged Britain's development into a modern industrialised society (Briggs 1983: 185–249). As indicated above, an increase of education for the working classes, which produced more readers, was part of the process, and it effectuated the dissolution of class barriers on the literary market. Forces both from within the middle-class literary system and from below initiated that dissolution, and it was connected to book formats and contents alike. The first rival for the three-decker system and the circulating library was serial publication, starting in the late 1830s when Charles Dickens and a few others managed to sell their novels to the growing reading public not in three volumes but in monthly instalments. Dickens had achieved an unexpected success on the print market with his *Sketches by Boz* (1836) and the *Pickwick Papers* (1836–37), whose late instalments reached print runs of 40,000 copies (Sutherland 1988a: 184, Patten 1979; cf. Sutherland 1988b on the success of the instalment method generally). The episodes in *Sketches*, loosely connected episodes of amusing observations of contemporary social life, of course lend themselves well to serial publication. The effect of Dickens' use of that format must not be underestimated, since 'subsequent to Dickens's success with *Pickwick*, parts publication became for 30 years a chief means of democratizing and enormously expanding the Victorian book-reading and book-buying public' (Patten 1979: 45).

An instalment of 32 pages would cost a shilling and it usually contained two or three chapters of a novel. The individual issues were of course much more affordable than a three-decker novel (or the circulating library membership fee), even if a long-running series would amount to the price of an expensive three-volume book. The similarities between Dickens' instalments and the shilling shockers and chapbooks of the 'low' market are obvious, and it is not least due to the format that his novels were perhaps the first to be 'popular' in the sense that they managed to unite readers from all classes. Beside instalments sold individually, Dickens also published his novels in parts that appeared in weekly or monthly magazines, as did many other Victorian authors. The immense demand for novels allowed some very successful titles to hit the market in various formats anyway, and again it is Dickens who stands out as a remarkable success. As Sutherland (1976a: 33) points out, Dickens' *Oliver Twist* was first serialised in *Bentley's Magazine* (February 1837–April 1839), but it was also published in one volume, in three volumes, as a reprint in ten instalments and again as a one-volume reprint in the subsequent eight years. Dickens' novels, however, did not entirely reach the 'lowest' market segments, though there was an appetite for his stories in that segment, too. They were even popularised further: *Pickwick Papers* were plagiarised for a cheaper, and shorter, series titled *Penny Pickwick*, as were many of his subsequent works, such as *Nicholas Nickleby*, pirated for *Nicholas Nickelberry*, *Oliver Twist* (*Oliver Twiss*) and *Martin Chuzzlewit* (*Martin Guzzlewit*), plundered by hack writers for Edward Lloyd, the publisher of penny dreadfuls (Altick 1957: 292; Anglo 1977: 76–78). Running over hundreds of numbers, these series of 'popularised Dickens' can justifiably be called 'marathon fiction' (Altick 1957: 292). Had the demand not existed from readers who found a Dickens novel too difficult, too long or too expensive, no penny versions would have been produced – in other words, a very 'low' market of fiction still existed at that time.

Among the forces that worked against the dominance of the three-decker was the simple demand for the books that people did not want, or could not afford, to buy in three volumes. Publishers who wished to sell such novels at affordable prices could reprint them in one volume and sell them at prices between 2s and 6s, which for most of the century could be done roughly a year after a novel's first publication in three volumes (Dooley 1992: 54–85; Eliot 1985). Two major examples of cheaper reprints are book series made available by W.H. Smith and by the publishing house of Bentley and Son. W.H. Smith, the leading bookseller and newsvendor company, opened bookstalls in the stations of the rapidly growing railway network that covered Britain almost totally between the 1840s and the 1870s. Starting with the first shop in 1848, W.H. Smith maintained shops on all main lines by 1860. In the 1850s, they sold one-volume reprints at 2s to travellers who wished to shorten the time of their journeys by reading: '[H]andily sized, these volumes were embellished by pictorial covers and are the forerunners of the twentieth-century paperback' (Sutherland 1988a: 589). Smith also established a circulating library system for train travellers, which grew to be the second force on the market after Mudie (Griest 1970: 31ff.; Sutherland 1976a, pp. 30ff.). Henry Colburn and Richard Bentley, of Bentley and Sons, published their Standard Novels series from 1831 on, which offered original fiction that had first appeared in three volumes in one volume at the price of 5s or 6s (Altick 1957: 274). Towards the 1880s, the period of time that elapsed between a novel's appearance as a three-decker and its cheap reprint was undercut more often than not, which eventually led to the demise of the three-volume format.

Before that happened, the 1860s saw the birth of a new sub-genre of prose fiction, the sensation novel, much discussed for its potential threats to morality and aesthetics (Maunder 2004). Mrs Henry Wood's *East Lynne* (1860) and Wilkie Collins' novels are among that group, as are Mary Elizabeth Braddon's bestsellers (e.g. *Lady Audley's Secret*, 1861, and *Aurora Floyd*, 1862). These novels deal with domestic crimes threatening the home, property and marriage. Unfaithfulness and bigamy, theft, embezzlement, fraud, arson, etc. jeopardise the mainstays of the Victorian middle-class ideology of domesticity (Hughes 1980, Maunder 2004 and 2005; see Harrison & Fantina 2006 for a view on sensation elements to be found in more novels than the core group, and also before the 1860s). Such novels toyed around with the notion that the security of the Victorian mentality might rest on an unsound footing, though they tended to establish bourgeois normality again towards the end – the thrill of insecurity was thus contained within the framework of middle-class value systems. Many of them were first serialised in journals that ranged from cheap weekly 1d papers to monthly 1s magazines, but also published in three volumes immediately after the final serial number, and thus made available to the downmarket buyers as well as the circulating library borrowers. That middle-class readers were attracted 'downwards' by sensation novels was a development that many a middle-class critic observed with dismay, including authors who saw their own efforts undervalued by the larger reading public. In their criticism, observers frequently used metaphors that linked reading sensation fiction with illicit sexual pleasures, over-indulgence in food (and poisonous food, at that) or intoxication (Gilbert 2003). The sensation novel was clearly an assault on the divisions between the three-decker and the popular segments of the literary system, blurring the distinctions between shocking and serious.

The last two decades of the nineteenth century saw the final blow to that constellation and a further fusion of low- and middle-class markets into the new popular, or middlebrow, novel. Not only did the three-decker die out: excitement continued its triumph among readers of all classes. Writers such as Henry Rider Haggard (*King Solomon's Mines*, 1885 and *She*, 1887) wrote sensational and imperialistic adventure stories for 'all the big and little boys who read it', as the dedication of *King Solomon's Mines* has it. Magazines featuring exciting stories for young readers flourished, attracting not only children but also older readers with an appetite for adventure, too. Robert Louis Stevenson (*Dr Jekyll and Mr Hyde*, 1886) and Bram Stoker (*Dracula*, 1897) revived elements of gothic horror in their narratives. These novels catered to the educated middle classes and the beneficiaries of the educational reforms alike, and both groups found as much entertainment and food for thought in them as they wished. Middle-class authors with an interest in the novel as a form of serious art – starting with George Eliot in the 1860s to 1880s, and followed by Thomas Hardy, George Meredith and George Moore, to name only a few late-Victorian authors – attempted to keep an educated middle-class readership interested in thematic and aesthetic innovation. They scared away the broad readership, however, which was increasingly unwilling to follow their experiments. More and more readers turned to the more easily digestible fictions that can, from that historical phase on, be called popular by way of distinction from 'artistic' (cf. Keating 1989: 436f.). The artistic novel turned into an elite art form followed only by small circles of highly educated readers, while novelists aiming at the broader markets at the middle and lower strata of society stuck to conventional aesthetics and ethics, telling stories with a focus on plot and action, without necessarily having to focus on horror and crime any longer. It took until the 1910s for an avant-garde movement – modernism – to emerge on the basis of that division, appreciated by the connoisseur, but not popular with the majority of readers.

References

Altick, Richard D. 1957. *The English Common Reader: A Social History of the Mass Reading Public 1800–1900*. Chicago, IL, London: Chicago University Press.

Altick, Richard D. 1973. *Victorian People and Ideas: A Companion for the Modern Reader of Victorian Literature*. New York: Norton.

Altick, Richard D. 1989. *Writers, Readers, and Occasions: Selected Essays on Victorian Literature and Life*. Columbus: Ohio State University Press.

Anglo, Michael. 1977. *Penny Dreadfuls and Other Victorian Horrors*. London: Jupiter.

Banham, Rob. 2007. 'The Industrialization of the Book.' Simon Eliot and Jonathan Rose (eds.). *A Companion to the History of the Book*. Oxford: Blackwell, pp. 273–290.

Bourdieu, Pierre. 1984. *Distinction: A Social Critique of the Judgment of Taste*. Trans. Richard Nice. Cambridge, MA: Harvard University Press.

Brantlinger, Patrick. 1998. *The Reading Lesson: The Threat of Mass Literacy in Nineteenth-Century British Fiction*. Bloomington, IN: Indiana University Press.

Briggs, Asa. 1983. *A Social History of England*. London: Penguin.

Dooley, Allan C. 1992. *Author and Printer in Victorian England*. Charlottesville, London: University Press of Virginia.

Eliot, Simon. 1985. 'The Three-Decker Novel and Its First Cheap Reprint 1862–1894.' *The Library* 7.1, pp. 38–53.

Eliot, Simon. 1994. *Some Patterns and Trends in British Publishing 1800–1919*. Occasional Papers of the Bibliographical Society, vol. 8. London: The Bibliographical Society.

Eliot, Simon. 2007. 'From Few and Expensive to Many and Cheap: The British Book Market 1800–1890.' Simon Eliot and Jonathan Rose (eds). *A Companion to the History of the Book*. Oxford: Blackwell, pp. 291–302.

Erickson, Lee. 1996. *The Economy of Literary Form: English Literature and the Industrialization of English Publishing, 1800–1850*. Baltimore, London: Johns Hopkins University Press.

Feather, John. 1988. *A History of British Publishing*. London, New York: Routledge.

Flanders, Judith [n.d.]. 'Penny dreadfuls.' London: The British Library. <http://www.bl.uk/romantics-and-victorians/articles/penny-dreadfuls> (24 June 2015).

Gilbert, Pamela K. 2003. 'Ingestion, Contagion, Seduction: Victorian Metaphors of Reading.' Tamar Heller and Patricia Moran (eds.). *Scenes of the Apple: Food and the Female Body in Nineteenth- and Twentieth-Century Women's Writing*. New York: State University of New York Press, pp. 65–86.

Golby, J.M. and A.W. Purdue. 1999 [1984]. *The Civilisation of the Crowd: Popular Culture in England 1750–1900*. 2nd ed. Stroud: Sutton.

Griest, Guinevere L. 1970. *Mudie's Circulating Library and the Victorian Novel*. Bloomington, IN: Indiana University Press.

Harrison, Kimberley and Richard Fantina. 2006. *Victorian Sensations: Essays on a Scandalous Genre*. Columbus, OH: Ohio State University Press.

Hewitt, Martin. 2013. *The Dawn of the Cheap Press in Victorian Britain: The End of the 'Taxes on Knowledge', 1849–1869*. London: Bloomsbury.

Hughes, Winifred. 1980. *The Maniac in the Cellar: Sensation Novels of the 1860s*. Princeton, NJ: Princeton University Press.

James, Louis. 1963. *Fiction for the Working Man 1830–1850*. London: Oxford University Press.

James, Louis. 2006. *The Victorian Novel*. Oxford: Blackwell.

Keating, Peter. 1989. *The Haunted Study: A Social History of the English Novel 1875–1914*. London: Secker & Warburg.

Kid, Allan and David Nicholls (eds.). 1998. *The Making of the British Middle Class? Studies of Regional and Cultural Diversity since the Eighteenth Century*. Stroud: Sutton.

McKitterick, David (ed.). 2009. *The Cambridge History of the Book in Britain, vol. vi: 1830–1914*. Cambridge: Cambridge University Press.

Maunder, Andrew. 2004. *Sensationalism and the Sensation Debate*. London: Pickering and Chatto.

Maunder, Andrew. 2005. 'Mapping the Victorian Sensation Novel: Some Recent and Future Trends.' *Literature Compass* 2.1, pp. 1–33.

Mitch, David F. 1992. *The Rise of Popular Literacy in Victorian England: The Influence of Private Choice and Public Policy*. Philadelphia: University of Pennsylvania Press.

Nelson, Robert L. 2005. 'The Price of Bread: Poverty, Purchasing Power, and The Victorian Laborer's Standard of Living.' *The Victorian Web: Literature, History and Culture in the Age of Victoria* <http://www.victorianweb.org/history/work/nelson1.html> (10 June 2015).

Neuburg, Victor. 1977. *Popular Literature: A History and Guide*. Harmondsworth: Penguin.

Patten, Robert L. 1979. *Charles Dickens and His Publishers*. Oxford: Clarendon Press.

Powell, Sally. 2004. 'Black Markets and Cadaverous Pies: The Corpse, Urban Trade and Industrial Consumption in the Penny Blood.' Andrew Maunder and Grace Moore (eds.). *Victorian Crime, Madness and Sensation*. Aldershot: Ashgate, pp. 45–58.

Rose, Mark. 1993. *Authors and Owners: The Invention of Copyright*. Cambridge, MA: Harvard University Press.

Springhall, John. 1994. '"Pernicious Reading"? The Penny Dreadful as Scapegoat for Late-Victorian Juvenile Crime.' *Victorian Periodicals Review* 27.4, pp. 326–349.

Storey, John. 2015. *Cultural Theory and Popular Culture: An Introduction*. 7th edition, London: Routledge.

Sutherland, John. 1976a. *Victorian Novelists and Publishers*. London: Athlone.

Sutherland, John. 1976b. 'The Economics of the Victorian Three-Volume Novel.' *Business Archives* 41, pp. 25–30.

Sutherland, John. 1988a. *The Longman Companion to Victorian Fiction*. Harlow: Longman.

Sutherland, John. 1988b. 'Chips off the Block: Dickens's Serialising Imitators.' Joanne Shattock (ed.). *Dickens and Other Victorians: Essays in Honour of Philip Collins*. London: Macmillan, pp. 97–119.

Suvin, Darko. 1982. 'The Social Addressees of Victorian Fiction: A Preliminary Enquiry.' *Literature and History* 8.1, pp. 11–40.

Thompson, Nicola Diane. 1996. *Reviewing Sex: Gender and the Reception of Victorian Novels*. Basingstoke: Macmillan.

Thompson, E.P. (ed.). 2011 [1975]. *Albion's Fatal Tree: Crime and Society in Eighteenth Century England*. Rev. ed. London: Verso.

Terry, R.C. 1983. *Victorian Popular Fiction, 1860–80*. London: Macmillan.

Vincent, David. 1983. 'Reading in the Working Class Home.' John K. Walton and James Walvin (eds.). *Leisure in Britain 1780 -1939*. Manchester: Manchester University Press, pp. 207–226.

Vincent, David. 1989. *Literacy and Popular Culture: England 1750–1914*. Cambridge: Cambridge University Press.

Walton, John K. and James Walvin (eds.). 1983. *Leisure in Britain 1780–1939*. Manchester: Manchester University Press, pp. 207–226.

Watt, William W. 1967. *Shilling Shockers of the Gothic School: A Study of Chapbook Gothic Romances*. New York: Russell and Russell.

Weedon, Alexis. 1996. 'Watch This Space: Wilkie Collins and New Strategies in Victorian Publishing in the 1890s.' Ruth Robbins and Julian Wolfreys (eds.). *Victorian Identities: Social and Cultural Formations in Nineteenth-Century Literature*. Basingstoke: Macmillan, pp. 163–183.

9 Picturing adventure

Popular fiction, illustration, and the British Empire, 1875–1914

Ralph Crane and Lisa Fletcher

A review of Charles G. Harper's 1892 book *English Pen Artists of To-day* in the *Illustrated London News* begins: 'The rapidly increasing recognition of illustration as the necessary accompaniment of both books and newspapers has brought the "pen artists" rapidly to the front' ('Pen artists', 1892: 261). Henry Blackburn (1896: n. pag.), in his *The Art of Illustration*, made a more emphatic claim for the increasing significance of illustration in the closing years of the nineteenth century: 'There is almost a revolution in illustration at the present time, and both old and young—teachers and scholars—are in want of a handbook for reference when turning to the new methods'. Illustration dominated late-nineteenth century popular culture, but it was by no means a stable sector of the publishing industry. As Richard Maxwell (2002: xxii) explains, during the nineteenth century, 'the illustrated book became a cultural form of the first importance' (see also Janzen Kooistra, 1995: 1; Buchholz, 2014: 593). By the century's final decade—as the Victorians' 'hunger for visual stimulation' (Maxwell, 2002: xxv) was intensified by photographic and cinematic technologies—illustration was ever-present but the *extent* of its dominance was waning. In particular, the illustration of popular novels, as a professional practice and as a cultural form, was in a state of high flux; in relation to novels, the 'revolution' Blackburn observed was a flurry of activity as illustration was nearing the end of its 'golden age'. Importantly, it was during this period also that the distinction between 'high' literature and popular fiction split the broader literary field, and novels began to be produced, circulated and 'classed' in relation to the system of genre which has defined and organized popular fiction ever since (Glover and McCracken, 2012; Gelder, 2004). This chapter turns to one genre that reached its apotheosis in the period known as British high imperialism—the juvenile adventure novel—to argue that the making of British popular literary culture was not solely a textual enterprise, and that our view of this genre's history is distorted to the extent that we ignore its visual elements.

David Glover (2012: 17) writes of 'the transformation of publishing in the period between 1880 and 1914 and its impact upon popular literature in the twentieth and twenty-first centuries, particularly through the inception of the modern system of genre fiction'. This system, still evident in the organization of bookshops in the English-speaking world has, for the most part, been studied as a literary or word-based phenomenon. There are a number of reasons for the 'cursory recognition' (Janzen Kooistra, 1995: 1) granted illustration in nineteenth-century literary histories, not least of which is the relative inaccessibility of the material. For example, undergraduate students who encounter adventure fiction of this period are most likely to read non-illustrated paperback editions of 'classic' novels and short stories by authors such as Arthur Conan Doyle, Rudyard Kipling, or Robert Louis Stevenson. They will rarely encounter the work of authors who, while household names in their own time, have largely faded into the archives except in the minds of postgraduate and academic researchers: G. A. Henty, W. H. G. Kingston, George Manville Fenn, and so on. As is widely known, the stories written by these men were published during a period of extraordinary transformation in popular literary culture—both in Britain and throughout the English-speaking world. Importantly, however, with rare exception their stories were first published, whether in serial or book form, with illustrations. Indeed, one of the lines of connection between the writers listed above is that the same artists often illustrated their work. For Paul Goldman (2012: 15), the 'breadth and depth' of nineteenth-century illustration that 'exists and remains relatively unexplored is staggering'; to date, the bias of most studies of the illustration of novels has been towards 'high' or canonical literature. From the perspective of popular culture studies—which recognizes that fields of visual and textual production overlap and cross-fertilise—there is enormous potential for new and exciting research in illustration: what does it mean, for example, that the pictures which were an essential component of popular novels were drawn by artists whose work was ubiquitous throughout Victorian culture? Book illustrators exhibited at the Royal Academy and in private galleries, worked as news illustrators for the popular press, painted full-colour lift-outs for magazines, and created images for (or which were adopted by) advertising campaigns. Due in part to websites such as *Internet Archive* and *Project Gutenberg*, and to the digitization projects of a number of national and university libraries, researchers now have free and ready access to an ever-increasing catalogue of digital scans of long out-of-print books and of the illustrated first editions of much-studied titles. This chapter looks at one small slice of an immense archive to show that, despite (or perhaps because of) the 'demanding interdisciplinary nature of the skills required to engage with the material' (Goldman, 2012: 18), paying more critical attention to illustration is essential to expanding our understanding of the place of fiction in the history of popular culture.

At the very least, as David Skilton (1988: 305) observes: 'The fact of illustration itself affects the sheer mechanics of the reading of fiction'. As we hope to show through detailed analyses of illustrations from three juvenile adventure fictions, the contributions made by artists to novels were multiple, frequently subtle, and sometimes surprising.

Simon Houfe (1978: 11) defines a book illustration as 'any pictorial subject in topography, architecture, genre or literature which aids a text, however slender'. In relation to imperial adventure fiction, arguments for the equivalency of words and images in illustrated books, such as Catherine J. Golden's (2000: 8) idea of 'picture-word collaboration', or Janzen Kooistra's (1995: 1) 'bitextuality', do not provide a helpful critical framework. If, as Gerard Curtis (2002: 1) insists, 'nineteenth-century literary culture in fact drew on a partnership between the textual and visual', then, certainly in relation to adventure novels, this partnership was a resolutely hierarchical one. The illustrator's job followed the work of the writer both sequentially and in terms of its priority or value: not only was the illustrator typically commissioned after the completion of the text, but the placement of their name in books and epitexts (including advertisements and reviews) was invariably secondary, and there was no guarantee that later editions would retain the initial illustrations. Paul Goldman and Simon Cooke (2012: 1) point to the period which illustration historians call 'the Sixties' (roughly 1855 to the mid 1870s) as one of 'radical change' in book production, both for developments in the 'aesthetics of the printed page', fuelled by major improvements in the technologies of illustration, and for the reconceptualization of illustration as 'a visual idiom in which the primary emphasis is on visualizing and interpreting the written word (Goldman and Cooke, 2012: 5). This view of the illustrator's role—to enhance the publication by illuminating or emphasizing some aspect of the narrative already communicated by the letterpress—held sway in the sector of fiction publishing discussed in this chapter.

British high (or new) imperialism was a phase of empire marked by vigorous colonial expansion and concomitant confidence, though not free from anxiety. Eric Hobsbawm usefully sets the parameters of the period in the title of his 1987 study, *The Age of Empire, 1875–1914*. During this period, the British Raj in India was at its height. The shock of the Uprising of 1857 and the distrust which followed in its aftermath had, 20 years later, given way to what Allen J. Greenberger (1969) terms the 'era of confidence' (1880–1910)—though as Kim A. Wagner (2010: 241) notes, the 'memory [of 1857] remained alive until the last days of the Raj'. Thus, in the last quarter of the nineteenth century, there was real impetus for juvenile popular adventure fiction to inculcate in readers a belief in the British Empire.

The imperial adventure novel, made popular by G. A. Henty, George Manville Fenn, and others, populated by manly heroes and set in hostile territory (India and elsewhere) was a perfect vehicle for promoting Empire. As Bradley Deane (2014: 1) explains: 'Men made the Empire, according to countless stories consumed by late Victorian and Edwardian readers, and, according to other stories just as numerous, the Empire made men'. Without disagreeing with Deane, we would add the codicil that the countless adventure fictions consumed by young readers at the end of the century knowingly sought to make the men who made the Empire.

These stirring tales were tools of popular imperialism through which young readers were introduced to the ideologies of masculinity and imperialism that drove British colonial expansion in what Mary Louise Pratt (1992: 4) calls 'contact zones' during high imperialism. However, imperial adventure was not solely a textual enterprise. Illustrations, as well as the letterpress they decorated and interpreted, were suffused with hegemonic ideologies of masculinity and imperialism. Illustrators, like writers and editors, were intent on selling the idea of Empire, and illustrations in periodicals and books made meaning in tandem with the texts they accompanied (see Crane and Fletcher, 2012).

The visual archive of Britain's relationship with India in the nineteenth century is unified by remarkably consistent conventions. Especially in its later decades, popular artists and illustrators routinely drew on a set of established tropes for portraying adventure in India. These conventions had much to do with prevailing ideas about India and Indians and, related to this, a powerful and seemingly unquestioned belief in British racial superiority.

As John Springhall acknowledges in his brief study of some of the illustrators of Henty's novels, it is difficult to find information about popular illustrators in the late nineteenth century. While we do not go so far as Springhall (1988: 17), who claims 'it should be evident, nonetheless, that the illustrations supplied for Henty's books by such artists merit equal importance with the text', we agree with his position that the work of illustrators needs further scholarly attention 'if we are to understand the process whereby romantic heroism, manliness and British imperialism were projected in an entertaining and escapist form to the schoolboy audience for whom they were chiefly intended'. Stuart Sillars (1995: 1) poses pertinent questions which guide his reading of illustrations:

> What exactly does a picture do in a book? How does it modify the response of the reader to the written word? Does it enable him or her to enter more fully into the created fictive world, or does it present that world as more of an artifice? Does it amplify the concepts and structures of the words, or does it offer separate ones of its own, as a sort of visual commentary? If the two work together to provide a new discourse, how is it assimilated and how best analysed?

Following Sillars' lead, this chapter offers close analyses of exemplary illustrations that focus on their inseparable narrative and ideological functions. Further, it is informed by appreciation of the cultural, social, and industrial context within which adventure novels were produced.

This chapter focuses on a selection of illustrations from three juvenile adventure novels set in India: G. A. Henty's *With Clive in India; or, The Beginnings of an Empire,* illustrated by Gordon Browne (1884); George Manville Fenn's *Gil the Gunner; or, The Youngest Officer in the East*, illustrated by W. H. Overend (1892); and Frederick P. Gibbon's *The Disputed*

V.C.: A Tale of the Indian Mutiny, illustrated by Stanley L. Wood (1904). In historical terms, these novels cover two 'critical events', not just in the history of British India but in the history of the British Empire as a whole: the Battle of Plassey in 1757, and what, in Victorian Britain (and in Anglo-Indian adventure fiction of the period), was commonly referred to as the Indian Mutiny of 1857.

Gordon Browne (1858–1932) was the son of the painter and engraver Hablot Knight Browne (1815–1882), better known as 'Phiz', and famous for his collaboration with Charles Dickens. While the former has attracted little scholarly attention, the few, mostly bio-bibliographical, sources which do discuss him seem to have reached a critical consensus: the son was a better trained and thus 'superior' artist to the father, but he 'never reached the latter's stature' (Houfe, 1978: 247). For Brian Doyle (1968: 315), Browne was 'one of Britain's most brilliant and prolific illustrators'. Richard Dalby (1991: 33) agrees, commending Browne as 'one of the greatest illustrators of the Golden Age [of children's book illustration], both in terms of quality and quantity'. Kate Jackson (2001: 111) lists Browne as one of the black-and-white artists who contributed to *The Strand* and 'became household names', thus acknowledging his place at the forefront of late Victorian popular culture. Why, then, have scholars paid so little attention to Gordon Browne? Dalby (1991: 33) speculates that his prolificacy, averaging six books a year for 50 years, may have detracted from his reputation as a significant artist. Brigid Peppin and Lucy Micklethwait (1984: 60) suggest further that the lack of critical acclaim accorded to Browne is 'due partly to the fact that he never developed a long-term association with an author of note (as did his father with Charles Dickens), and partly because much of his vast output was published in very cheap editions' (see also Houfe 1978: 247). There is no doubt that 'GB' has not endured as a signifier of a 'Golden Age' of illustration to anything like the same extent as 'Phiz'—and for some compelling reasons. However, the scholarly agreement about Browne's achievements has not been informed by careful study of his prodigious output or its reception during his lifetime.

With Clive in India won high praise from reviewers in the popular press when Blackie and Son published it for the 1883 Christmas season. In a typical review of the new titles for the season, the *Bristol Mercury and Daily Post* commends the 'admirably-got-up Christmas gift-books of Messrs. Blackie and Son, who apparently spare neither pains nor expense in the production of this class of literature' ('Christmas Books', 1883: 6). Featuring 12 full-page line-and-wash drawings by Browne, *With Clive In India* was one of 13 titles advertised widely by Blackie as 'New Books for the Young' from November 1883. At six shillings it was one of Blackie's two most expensive gift-books that year; the other was George Manville Fenn's *The Golden Magnet: A Tale of the Land of the Incas*, which also included 12 drawings by Browne. Browne illustrated over half of Blackie's 1883 Christmas volumes. There is wide critical agreement that Henty 'dominated children's literature between 1871 and 1906' (McMahon, 2010: 154), but when juvenile fiction of the period is acknowledged as an *illustrated* form, it becomes significant that the name mentioned most often in the publisher's Christmas advertising campaign in 1883 was not Henty's but Browne's.

In December 1883, 'The Fairyland Review', a satirical poem spoofing the announcements and reviews of 'Christmas Books' which were ubiquitous in newspapers and magazines of the period, appeared in *Punch*. The penultimate verse focuses on Browne, whose illustrations link the several authors listed:

> *Sheer Pluck, With Clive in India*, are books boys *can't* put down,
> The author is GEORGE HENTY, and the artist GORDON BROWNE;
> Who, in *The Golden Magnet*—by the skilful MANVILLE FENN—
> With clever graphic pencil gives a piquancy to pen:
> In *Wigwam and the War-path*, too, his talent has full scope,
> To illustrate the stories that are told by ASCOTT HOPE.
> *Picked up at Sea*, by HUTCHESON, you'll gladly read, no doubt,
> And a score or two of others we can't now write about.
>
> ('The Fairyland Review', 1883: 273)

Henty and Browne are again paired in *Punch* in a similar verse the following year: 'GEORGE HENTY, also GORDON BROWNE, can't fail to win applause' ('Paper-Knife Poems', 1884: 274). In 1887, a long stanza entitled 'Messrs. Blackie & Sons' in 'The Babes in the Christmas Wood: A Publisher's Cantata' (1887: 266), begins with Henty—one of the 'good names' who authors 'tales of home life and adventure in plenty'—and closes with 'GORDON BROWNE, whose famed pencil so skilled, / Of the foregoing pages so many has filled'. Illustrations, especially by a 'famed pencil', were clearly a selling point for publishers of juvenile fiction throughout the 1880s, but this trio of verses suggests that the pictures were not incidental to the cultural or, indeed, ideological work of the books in which they appeared.

A review published in *The Academy* begins with a statement of high praise for the two men named on the novel's title page:

> Among writers of stories of adventure for boys, Mr. Henty stands in the very front rank; and Mr. Gordon Browne occupies a similar place with his pencil. It is due also to Messrs. Blackie to add that no firm of publishers turns out this class of literature with more finish. We refer not only to the novel tinting of the illustrations and the richness of the cover, but also more particularly to the solidity of the binding—a matter of great importance in boys' books. Of the story before us it is not necessary to say very much. [...] [O]ur final verdict is that those who know something about India will be the most ready to thank Mr. Henty for giving them this instructive volume to place in the hands of their children.
>
> ('Gift-Books', 1883: 362)

The Academy comes very close to presenting *With Clive in India*—and 'adventure for boys' more broadly—as a collaborative enterprise involving author, illustrator, publishers, and printers. The review's 'final verdict', however, points to the powerful hierarchy in high imperial juvenile publishing, which positioned authors at the apex of book culture: thanks are due to Mr. Henty for this 'instructive volume'.

With Clive in India sold well; first released on 24 September 1883, sales reached 2,773 copies by the end of February 1884, and averaged 1,008 copies annually for the next 15 years (Newbolt, 2005: 96). In August 1906, Blackie released a 'cheap edition' which sold almost 20,000 copies in ten years. It includes six full-page 'wash-drawings by an unknown hand' based on Browne's original illustrations (Newbolt, 2005: 99), who by this time was attached to rival publisher Cassell (Darton 1899: 69). This novel's publication history—and it is not a peculiar one—is a caution against overstating the 'authorial' claim of illustrators. Browne was unequivocal about the relative status of author and artist in the production of illustrated books:

> The black-and-white artist must illustrate the story; to do that properly, he must read it—really read it—and try to appreciate the feelings of the author. His own impressions may be very good, but the author's are those that are required.
>
> ('An Illustrious Illustrator', 1894: 535)

The frontispiece illustration may have been the first thing Victorian readers saw after opening a novel, but the proximity of this image to the title page means that any speculations about the value and significance of the work of men like Browne must take into account the compelling paratextual evidence that their work was considered supplementary to the written text. Further, the inclusion of copies of Browne's images by an unknown illustrator in the 1906 edition points to the waning significance of black-and-white artists to book publishing in the first decade of the twentieth century—which paradoxically may indicate their importance in laying the groundwork for the emergence of a hyper-visual popular culture in the twentieth century.

Browne's illustrations for *With Clive in India* were completed relatively early in his long career, but at a time when his reputation was already firmly established. R. E. D. Sketchley (1903: 96) positions Browne 'at the forefront of illustrators popular with story-book publishers and with readers of story-books' since the 1880s. Sketchley (1903: 96) finds in Browne's work 'no trace of the "caricaturizations" of "Phiz"'. Curiously, however, one of his illustrations for *With Clive in India* (Figure 9.1) does seem to gesture to 'Phiz', in both its composition and the strong 'trace of 'caricaturization'. 'How Charlie Dealt with his Superior Officers' pictures the novel's 'honourable and manly' (Henty, 1884: 9) hero, Charlie Marryat, in the final moments of a dramatic confrontation with three of the men who ignominiously deserted Fort William when it was besieged by the

Figure 9.1 'How Charlie Dealt with his Superior Officers'. Illustration by Gordon Browne. G. A. Henty, *With Clive in India* (London: Blackie, 1884), facing 230.

Nawab of Bengal, Siraj-ud-daulah, in 1756: the governor, Roger Drake; the garrison commander, George Minchin; and Andrew Young, the captain of the *Dodaldy*, the ship on which Drake, Minchin, and others fled downriver, leaving behind those who would later be imprisoned in the infamous 'Black Hole' of Calcutta. Charlie is an anachronistic character who shares the avuncular narrator's late-Victorian perspective on Clive and the historical events in which he becomes embroiled. In this episode, he volunteers to swim to the ships under cover of night and pass on orders for the deserters to return. Charlie exclaims to the three men depicted—the most Dickensian figures in the book's illustrations—that 'England will not believe that three Englishmen so cowardly, so infamous as yourselves, are to be found', then 'with all his strength' he strikes the governor between the eyes (ibid: 231). In Browne's drawing Drake is sprawled beneath a well-set table, his powdered wig awry; Michin and Young, buttons straining over their well-fed bellies, appear unbalanced by Charlie's effrontery, jowly and puff-cheeked with indignation; while Charlie stands tall and straight, his wide-legged stance and broad shoulders providing the reader a stable point of identification.

With Clive in India is a dense account of roughly ten critical years, from the mid-1750s to the mid-1760s, encompassing the Third Carnatic War (1756–1763) and the Battle of Plassey (1757), which together saw the British cement their rule in Bengal and much of southern India and lay the groundwork for the British Raj. As Henty (1884: v) acknowledges in his preface, the novel gives 'a somewhat smaller space to the personal adventures of my hero than in my other historical tales', focusing instead on historical details culled from

Figure 9.2 'The Bazaars of Trichinopoli'. Illustration by Gordon Browne. G. A. Henty, *With Clive in India* (London: Blackie, 1884), facing 118.

Robert Orme's *A History of the Military Transactions of the British Nation in Indostan* and G. B. Malleson's biography of Clive. However, 11 of the 12 illustrations focus on Charlie's adventures in India, a point of contrast which suggests one of the ways in which pictures enhanced the reading experience for the 'Dear Lads' to whom Henty (ibid: v) addresses the preface and promises 'thrilling and exciting' events: by maintaining a focus on the hero even when he retreats to the background of the written text. The subject and placement of Browne's fifth illustration 'The Bazaars of Trichinopoli' (Figure 9.2) invites readers to study the connections between the letterpress and the book's visual elements.

Charlie and his friend Peters, by this time East India Company soldiers, arrive in Trichinopoli in late 1753, shortly after the conclusion of the siege (1751–1752) and explore the town, like late-Victorian tourists taking in key sites: 'After visiting the fort they rambled through the town, and were amused at the scene of bustle in its streets and at the gay shops, full of articles new and curious to them, in the bazaars' (ibid: 119). Charlie stands front and centre in Browne's interpretation of this busy scene, which, by offering readers a detailed wide-angle snapshot of a 'tour' of India, is reminiscent of nineteenth-century news illustration. This drawing's reality effect is further enhanced by the prosaic caption and the illustration's proximity to the half-page map, 'The Environs of Trichinopoli, to illustrate the Military Operations of 1751–1753' (ibid: 120), which concludes the chapter. Browne pictures Charlie at leisure, taking a moment during his ramble to deliver a short speech to Peters about the relative strength of Englishmen and Indians.

Pictured in the confident wide-legged stance so common of the British in high imperial illustrations, Charlie is fascinated by the 'wonderfully clever and ingenious' (ibid: 119) activity on display and expresses confusion that such capable artisans only become 'good fighting men' (ibid: 120) when commanded by Europeans. This simplistic justification for British rule in India is arguably the common theme of Browne's illustrations, which, like Henty's narrative, cast Charlie as a natural-born leader for a British-led India—a capacity not shared by his Irish friend and 'soldier servant' (ibid: 119) Tim Kelly.

Midway through the novel, Charlie, now fluent in the vernacular (the array of languages in southern India is conveniently glossed over), is commissioned by the governor of Madras, Thomas Saunders—on the personal recommendation of Clive—to travel in disguise to the fortress of the chief Boorhau Reo. Once there, Charlie is to 'train his men after the European fashion' (ibid: 135) so that the fictional rajah will be better able to defend his territory against the more powerful Marattas or the Nizam of Hyderabad. Charlie is such an effective military agent that he provokes the 'hostility' of the 'leading men of the state' (ibid: 151) and two attempts are made on his life. From the first, he gains a loyal servant when he shows mercy to the cook, Hossein, who had allowed Charlie's food to be poisoned. Hossein later raises the alarm when assassins attack Charlie in his bedroom. The most dramatic illustration in the novel depicts 'two of the conspirators' who, unable to evade capture, 'ran to the parapet and leaped over into the valley, two hundred feet below' (ibid: 162). With this illustration (Figure 9.3), Browne, who Springhall (1988: 11) notes had an 'inclination for figures falling through space', emphasizes an event which takes up only a sentence in Henty's text.

Browne ('An Illustrious Illustrator', 1894: 535) explained his process for selecting scenes for illustration:

> I read [the book] thoroughly once, putting a pencil mark against the passages which particularly strike me. Then I read it again, 'skipping' it this time, and weed out my marked passages. Sometimes, if a book pleases me very much, I read it again. Vivid, striking descriptions of stirring incidents lend themselves best to illustration.

Human figures are invariably the main elements of popular imperial illustrations. The key figures are almost always, to adapt cinematic terms, portrayed in long shot, their entire body visible within the frame of the image. The arrangement of figures in drawings such as 'Tim Kelly Discoursing on Horrible Reptiles' (Henty, 1884: facing 86) and 'How Charlie Dealt with his Superior Officers' (Figure 9.1) thus suggest a theatre stage and position readers as an audience to events in which they are not directly implicated. Other illustrations such as 'A Brush with Privateers' (ibid: frontispiece) maintain this theatrical staging, but the positioning of figures at the borders

Figure 9.3 'The Fate of the Assassins'. Illustration by Gordon Browne. G. A. Henty, *With Clive in India* (London: Blackie, 1884), facing 162.

of the image, with only part of their body visible, creates an open and more naturalistic composition. Browne's rendition of the fate of the assassins eschews this convention for theatrical organization of the image and, despite the portrait format, is better described as cinematic. The suggestion of motion in the two falling bodies is a subtle reminder, too, of the limitations of photography in this period, which was better suited to the panoramic scale of the image's background than to capturing a split second of rapid movement.

Whereas 'The Fate of the Assassins' looks forward to the dynamic visual potential of cinema, the novel's penultimate illustration looks back to the classical art traditions that would have been the basis of Browne's formal training at Heatherley's Art School in London. Like the majority of plates in *With Clive in India*, 'The Sudden Conversion of a Faithless Ally' (Figure 9.4) is in portrait, with the focus of the action in the middle third. The background is sketchy and provides only enough detail to denote the setting; darker shades and stronger lines are reserved for the main figures and their 'props'. The image has a clear pyramidal composition, enhanced by the line of the tent's drapery in the background. The use of classical rules of composition and perspective—apparent in every picture in *With Clive in India*—are central to the efficiency of popular fiction illustrations, both

Figure 9.4 'The Sudden Conversion of a Faithless Ally'. Illustration by Gordon Browne. G. A. Henty, *With Clive in India* (London: Blackie, 1884), facing 344.

from the perspective of a working black-and-white artist like Browne who produced a staggering number and variety of images every year, and because the success of popular illustrations depended, in part, on their immediate legibility. This picture depicts Charlie holding a pistol to the throat of Rajah Anandraz, captured before he is able to betray the British forces to the Nizam. Tim draws his sword to the rajah's left and Hossein is just visible in the background, holding up his sword arm to hail supporters—the figures having been transposed by Browne in order to bring the two uniformed soldiers together in the forefront of his composition. This illustration, which is strongly reminiscent of Delacroix's *Liberty Leading the People*, is an excellent example of Browne's familiarity with and use of canonical paintings—the pictorial citation a more subtle version of the kind of textual borrowing for which Henty is well known.

Like Browne, W. H. Overend (1851–1898) and Stanley L. Wood (1866–1928) were both prolific illustrators of imperial adventure fiction. Overend was a marine painter and illustrator whose work appeared in almost every issue of the *Illustrated London News* for three decades from 1872, as well as in the *Boy's Own Paper*, and *Chums* (Houfe, 1978: 406). Following Overend's death, M. H. Spielmann (1898: 428) wrote that 'as a draughtsman in black and white he for many years held his position among the first of the artist-journalists of whom this country is proud'. Wood's illustrations also appeared regularly in the *Boy's Own Paper* and *Chums*, as well as in numerous other magazines from *The Graphic* and *The Strand* to *Young England*.

Manville Fenn's *Gil the Gunner*, illustrated by Overend, and Frederick P. Gibbon's *The Disputed V.C.*, illustrated by Wood, are both good examples of Mutiny novels aimed at a juvenile readership. Gibbon's young hero, Ted Russell, takes part in almost every major campaign of the rebellion—in much the same way as the juvenile heroes of Henty's overtly jingoistic *In Times of Peril* (1881). Gautum Chakravarty (2005: 114) notes that *The Disputed V.C.* is 'a colonial adventure novel in the high imperial mood of Henty and others'. *Gil the Gunner* is rather harshly summed up by Hilda Gregg (1897: 225) in her review of 'The Indian Mutiny in Fiction' as 'a story with no lack of exciting incidents, but suffering from the drawback of a hero of a type peculiar, we believe, to Mr Fenn's tales for boys, who can only be fitly described by the epithet cantankerous'. In contrast, Amanda Lynn Phillips (2009: 44) praises Manville Fenn's 'lush description, thoroughly readable dialogue (at least in comparison to Henty), and fairly smooth transitions from scenes of the protagonist's private actions, thoughts and conversations to vibrant scenes of large-scale battle'. The novel follows the eponymous hero, Gil Vincent, on his journey from boyhood to manhood through the fiery cauldron of the Mutiny, and the friendships he forges with two older men—his compatriot, Captain Brace, and the Indian rebel leader, Ny Deen—which are 'more than tinged with homoeroticism' (Phillips ibid: 47), a dynamic which Overend illustrates (Figure 9.10).

In its ambitious engagement with so many iconic sites of the Mutiny and so many tropes of the Mutiny romance, Gibbon's novel provides Wood with almost infinite scope for his illustrations. Thus, assuming Wood has 'really read' the novel, as Browne insists an illustrator must do, paying attention to his choice of narrative moments to picture in the three black-and-white plates that accompany the text can only enhance our understanding of high imperial popular culture. In all three illustrations, Wood chooses moments of (successful) British action: Ted in the act of blowing up a magazine; Battye leading a charge; Tynan shooting a rebel. In other words, his illustrations picture heroic deeds of adventure, which as Astrid Erll (2009: 120) claims is 'a very appropriate kind of remembering in [novels] which [are] clearly meant to instil imperialist values and norms in [their] young readers'.

Wood's frontispiece illustration (Figure 9.5) depicts the novel's titular moment, freezing the young hero in the act of blowing up the Aurangapore magazine to prevent it falling into rebel hands—an act that will earn him a V.C. once the false claim to the action by the novel's villain, Tynan, has been cleared up. Nevertheless, Ted's actions, though undoubtedly courageous, lack the derring-do that is highlighted in many illustrations of the period, making the composition somewhat unusual. Ted is caught mid movement, awkwardly leaning forward. The arm which drops the candle is extended, his body braced, and his accoutrements, a sword and holstered pistol, follow

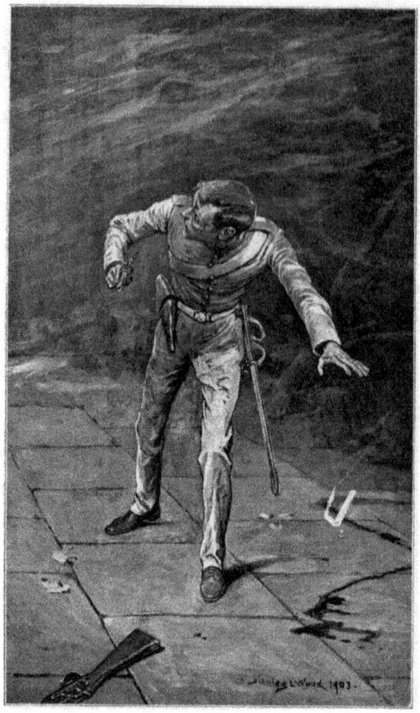

Figure 9.5 'He dropped the lighted candle on the thin trail of powder'. Illustration by Stanley L. Wood. Frederick P. Gibbon, *The Disputed V.C.* (London: Blackie, 1904), frontispiece.

the lines of his arms. Unusually, though, all point towards the ground: not in surrender, but to draw the reader's attention to the task in hand, while his averted face, glancing back over his shoulder, emphasizes the danger of his action. The image, while it contains only the single figure, is nevertheless an expansive one. The gaze back over his shoulder, the half-visible rifle, even the lines between the flagstones all make this a proleptic image, anticipating for the reader the 'explosive' action to come.

If Wood's frontispiece captures a critical moment, Overend's '"Mutiny," he panted; "Don't go back"' (Figure 9.6) represents a critical event. The background detail in this and other illustrations by Overend (in this text and elsewhere) is in marked contrast to the style of Browne, who favours indistinct backgrounds to focus the reader's gaze. Here, however, the reader is invited to read this busy, open image like a passage of text, taking in first the scene left and top before shifting their gaze to the wounded soldier in the bottom right, who points beyond the frame, to both the scene of the mutiny he has escaped, and, textually, to the central events of the narrative to follow.

Both Wood in 'Battye rose in his stirrups and thundered forth the order to charge' (Figure 9.7) and Overend in 'Dobbs blew forth a triumphant charge' (Figure 9.8) depict British soldiers galloping into battle, in what is

Figure 9.6 '"Mutiny," he panted; "Don't go back"'. Illustration by W. H. Overend. George Manville Fenn, *Gil the Gunner* (London: SPCK, 1892), facing 115.

perhaps one of the most ubiquitous of military adventure tropes. In Wood's illustration, a British soldier on horseback, his sword drawn and raised above his head charges into battle, while in Overend's image, though a bugler is forefront, the drawn and raised sword of an officer is visible on the left of the image. Compositionally, the focus of both images is divided between horse and rider. As the reader shifts their gaze in Wood's illustration between the strong lines of Lieutenant Battye's sword arm extended by his sabre to the horse's muscular body, they are drawn towards the soldier's face. Battye stands in his stirrups, looking back over his left shoulder towards the Guides who charge in his wake. The faces of the turbaned riders, particularly their eyes, suggest the dramatic intensity of the scene, which is mirrored in the faces of their horses. And, while Battye is the focal point, the energy and emotion of the illustration is contained in the muscular body of his horse, caught mid-gallop. This style of image was pervasive in Wood's work—the frontispiece to Henty's *Rujub, the Juggler* is another good example (see Crane and Fletcher, 2012: 155–56)—and in adventure illustration and military art more generally. Indeed, late-Victorian readers of novels illustrated by Wood would have been very familiar with the key thematic and stylistic elements of his work. A charging soldier on horseback was a staple, for example, of

Figure 9.7 'Battye rose in his stirrups and thundered forth the order to charge'. Illustration by Stanley L. Wood. Frederick P. Gibbon, *The Disputed V.C.* (London: Blackie, 1904), facing 160.

Figure 9.8 'Dobbs blew forth a triumphant charge'. Illustration by W. H. Overend. George Manville Fenn, *Gil the Gunner* (London: SPCK, 1892), facing 228.

academic battle painting and news illustration, and Wood's treatment of his subject is highly conventional, recalling well-known images by artists such as Lady Elizabeth Butler and Richard Caton Woodville.

In his final illustration for *The Disputed V.C.*, 'The rebel reeled against the wall' (Figure 9.9), Wood portrays a critical moment: the death of a rebel sepoy and the concurrent rescue of Ted Russell by his nemesis, Tynan, an erstwhile cad who redeems himself by this action. It delivers an instant of narrative crisis that will decide the course of events for the young hero. In the foreground a sepoy, or 'pandie' as Gibbon prefers to call the mutineers, collapses against a wall, a tulwar falling from his right arm, which points impotently towards the ground. In the background Tynan, braced for the firing of his pistol, offers in contrast a symbol of British martial superiority. At his feet lies the wounded hero, saved from imminent death.

The symmetry of Wood's illustration is carefully constructed. Tynan's legs and the near 45 degree angle of his body, which leans towards the fatally wounded sepoy, suggests forward momentum and is emblematic of imperial right and might. In contrast, the matching angle of the dying sepoy's body reveals him sliding towards the ground, his menace expunged. This whole composition is redolent of the representations of the deaths of both sepoys and tigers (as hunted prey and signifiers of the mutineers) in the visual archive of the late nineteenth century, and as such it delivers a full narrative of racial and cultural superiority on the page.

Figure 9.9 'The rebel reeled against the wall'. Illustration by Stanley L. Wood. Frederick P. Gibbon, *The Disputed V.C.* (London: Blackie, 1904), facing 336.

Figure 9.10 'We can be friends, and eat salt together'. Illustration by W.H. Overend. George Manville Fenn, *Gil the Gunner* (London: SPCK, 1892), facing 307.

An entirely different narrative appears in Overend's 'We can be friends, and eat salt together' (Figure 9.10). Rather than depicting another stirring moment from the adventure, this illustration, which strays from the genre's preoccupation with manly imperialist values, shows an intimate moment between the wounded and bed-ridden hero and the man who saved his life and now holds him captive: the rebel leader Ny Deen, the Maharajah of Ahdenpore. The rich furnishings of Gil's tent, including his ornate bed, are depicted in detail, which is at odds with the usual sketchy backgrounds found in 'action' illustrations. Similarly, Ny Deen's elaborate dress is given the level of close attention more often applied to British military uniforms. The reader's eye, however, is drawn to the spotlit face of Gil, who gazes intently into the eyes of Ny Deen. This image—with the two men holding hands—works in tandem with the letterpress to capture and preserve the latent, interracial homoeroticism that characterizes the relationship between the two men, with which Gil struggles in this and several similar scenes. This implicit homoeroticism continues in Overend's final illustration, 'I could not refrain from drawing the flashing blade' (the pun on 'drawing' is surely no accident), which pictures Gil dressed in the fine military uniform presented to him by Ny Deen, and caressing the blade of a bejewelled tulwar that is also a gift. In this portrait Gil, who only a short time before had made little effort to free his hand from his captor's grasp, appears to be accepting Ny Deen's proposal (to join him as his General). In these illustrations, Overend succinctly captures the temptations—military and sexual—Gil faces and overcomes.

It is no accident that all the images discussed in this chapter are exclusively masculine in content, depicting the imperial struggle as a contest between men. This seemingly uncomplicated representation of Empire is repeated in the majority of illustrated adventure fictions for boys published during the period of high imperialism.

All three illustrators worked within both the conventions of the popular adventure genre that reached its zenith in this period, and the recognized conventions of 'high' and 'low' art. Individually and collectively, the images they produced certainly affirmed the ideology of empire in the popular imagination, but more generally, in Britain and abroad, they fed the expectation that popular storytelling would be at once verbal and visual. As Patricia Anderson (1991: 2) argues with reference to the mid-nineteenth century, 'the hallmark of a transformed and expanded popular culture was its increasingly pictorial character'. By the early twentieth century, the pictorial character of popular culture was irrevocably established, and as the example of popular adventure fiction shows, book illustrators had played a major part in this dramatic development.

References

Anderson, P. (1991) *The printed image and the transformation of popular culture, 1790–1860*, Oxford: Clarendon Press.

'An illustrious illustrator: Afternoon tea with G.B.' (1894) *Chums*, vol 2, no. 84, p. 535.

Blackburn, H. (1896) *The art of illustration*, London: W. H. Allen & Co, viewed 22 May 2015, http://www.gutenberg.org/ebooks/32320

Buchholz, L. D. (2014) 'Illustrations and text: Storyworld space and the multimodality of serialized narrative', *Style*, vol. 48, no. 4, pp. 593–611, viewed 26 May 2015, ProQuest Central.

Chakravarty, G. (2005) *The Indian mutiny and the British imagination*, 'Christmas Books', *The Bristol Mercury and Daily Post* (17 November 1883), p. 6, viewed 16 June 2015, 19th Century British Newspapers.

Crane, R., and Fletcher, L. (2012) 'Picturing the empire in India: Illustrating Henty', *English Literature in Transition, 1880–1920*, vol. 55, no. 2, pp. 155–75.

Curtis, G. (2002) *Visual words: Art and the material book in Victorian England*, Aldershot: Ashgate.

Dalby, Richard. (1991) *The golden age of children's book illustration*, New York: Gallery Books.

Darton, J. W. (1899) 'An Illustrator of Books', *Art Journal*, pp. 69–72, viewed 22 May 2015, ProQuest British Periodicals.

Deane, B. (2014) *Masculinity and the new imperialism: Rewriting manhood in British popular literature, 1870–1914*, Cambridge: Cambridge University Press.

Doyle, B. (1968) *The who's who of children's literature*, London: Hugh Evelyn.

Erll, A. (2009) 'Remembering across Time, Space, and Cultures: Premeditation, Remediation and the "Indian Mutiny"', in Erll, A. and Rigney, A. (ed.) *Mediation, remediation, and the dynamics of cultural memory*, Berlin: Walter de Gruyter, pp. 109–38.

'The Fairyland Review', *Punch* (December 8 1883), p. 273, viewed 22 May 2015, Punch Historical Archive.

Gelder, K. (2004) *Popular fiction: The logics and practices of a literary field,* London: Routledge.

Gibbon, F. P. [1904] *The disputed V.C.: A tale of the Indian mutiny,* illustrated by S. L. Wood, London: Blackie and Son.

'Gift-Books', (1883) *The Academy*, vol. 604, p. 362, viewed 22 May 2015, Proquest British Periodicals.

Glover, D. (2012) 'Publishing, history, genre', in Glover, D. and McCracken, S. (eds.) *The Cambridge companion to popular fiction,* Cambridge: Cambridge University Press, pp. 15–32.

Glover, D. and McCracken, S. (ed.) (2012) *The Cambridge companion to popular fiction,* Cambridge: Cambridge University Press.

Golden, C. J. (2000) 'Introduction', in Golden, C. J. (ed.) *Book illustrated: Text, image, and culture, 1770–1930*, New Castle, DE: Oak Knoll Press, pp. 1–30.

Goldman, P. (2012) 'Defining illustration studies: Towards a new academic discipline', in Goldman, P. and Cooke, S. (ed.) *Reading Victorian illustration 1855–1875: Spoils of the lumber room,* Farnham: Ashgate, pp. 13–32.

Goldman, P., and Cooke, S. (eds.) (2012) *Reading Victorian illustration 1855–1875: Spoils of the lumber room,* Farnham: Ashgate.

Greenberger, A. J. (1969) *The British image of India: A study of the literature of imperialism*, London: Oxford University Press.

Gregg, H. (1897) 'The Indian Mutiny in Fiction', *Blackwood's Edinburgh Magazine* February, pp. 218–31.

Henty, G. A. (1884) *With Clive in India; or, The beginnings of an empire*, illustrated by G. Browne, London: Blackie and Son.

Houfe, S. (1978) *The dictionary of British book illustrators and caricaturists 1800–1914*, Woodbridge: Antique Collectors' Club.

Jackson, K. (2001) *George Newnes and the New Journalism in Britain, 1880–1910,* Aldershot, Ashgate.

Janzen Kooistra, L. (1995) *The artist as critic: Bitextuality in fin-de-siècle illustrated books*, Aldershot: Scolar.

McMahon, D. H. (2010) '"Quick, Ethel, your rifle!": Portable Britishness and flexible gender roles in G. A. Henty's books for boys', *Studies in the Novel*, vol. 42, pp. 154–72, viewed 23 November 2010, Project Muse.

Manville Fenn, G. (1892) *Gil the gunner; or, The youngest officer in the East*, illustrated by W. H. Overend, London: Society for Promoting Christian Knowledge.

Maxwell, R. (2002) 'Introduction', in Maxwell, R. (ed.) *The Victorian Illustrated Book*, Charlottesville: University Press of Virginia, pp. xxi-xxx.

Newbolt, P. (2005) *G. A. Henty 1832–1902: A bibliographical study of his British editions with short accounts of his publishers, illustrators and designers, and notes on production methods used for his books*, Philadelphia, PA: Polyglot Press.

'Paper-knife poems,' *Punch* (December 6 1884), p. 274, viewed 22 May 2015, Punch Historical Archive.

'Pen artists', review of *English pen artists of to-day: Examples of their work with some criticisms and appreciations*, by Charles G. Harper, *Illustrated London News* (27 February 1892), p. 261, viewed 24 May 2015, Illustrated London News.

Peppin, B. and Micklethwait, L. (1984) *Book illustrators of the twentieth century*, New York: Arco Publishing.

Phillips, A. (2009). *Masculinity in peril: Mutiny fiction and Victorian man-making*. M.A. University of Alabama, viewed 1 June 2015, http://acumen.lib.ua.edu/content/u0015/0000001/0000006/u0015_0000001_0000006.pdf

Pratt, M. L. (1992) *Imperial eyes: Travel writing and transculturation*, London: Routledge.

Sillars, S. (1995) *Visualisation in popular fiction 1860–1990: Graphic narratives, fictional images*, London, Routledge.

Sketchley, R. E. D. (1903) *English book-illustration of to-day: Appreciations of the work of living English illustrators with lists of the books*, London: Kegan Paul, viewed 16 April 2015, http://www.gutenberg.org/files/38164/38164-h/38164-h.htm

Skilton, D. (1988) 'The relation between illustration and text in the Victorian novel: A new perspective', in Höltgen, K. J. and Daly, P. M. (ed.) *Word and visual imagination: Studies in the interaction of English literature and the visual arts*, Erlangen: Universitätsbibliothek Erlangen-Nürnberg, pp. 303–25.

Spielmann, M. H. (1898) 'An artistic canserie', *The Graphic*, 2 April, p. 428, viewed 11 June 2015, British Newspapers 1650–1950.

Springhall, J. (1988) '"Dreams of adventure, deeds of empire": Some Henty illustrators reconsidered', *Henty Society Bulletin*, vol. 43, no. 6, pp. 10–20.

The babes in the Christmas wood: The publishers' cantata' (1887) *Punch* (10 December), p. 266, viewed 22 May 2015, Punch Historical Archive.

Wagner, K. A. (2010) *The great fear of 1857: Rumours, conspiracies and the making of the Indian uprising*, Oxford: Peter Lang.

10 'For the benefit of old boys, young boys, odd boys generally, and even girls'

The irresistible rise of the British comic, 1884–1900

Robert Shail

The period from 1884 to 1900 saw the appearance in Britain of a new form of mass entertainment: the comic. While other European countries, and the United States, developed their own comics around this time, the British version was distinctive with a character which reflected the specific cultural environment which produced it. Its arrival is often dated to the first issue of *Ally Sloper's Half Holiday* in 1884, whose tagline provides the title for this essay. It was rapidly followed by other titles such as *Comic Cuts* and *Chips*. These papers were linked by both their format and their appeal to a largely urban working-class readership. Cheaply priced and published weekly, they soon established an enormous audience. Their popularity took advantage of significant changes in the later part of the nineteenth century, including the development of printing technology and the impact of the 1870 Education Act. The latter greatly enhanced the growth of literacy among the industrial working-class who had, for the first time in large numbers, the ability to read and enjoy comics. A key feature of the new form was the way it responded to this core readership and began to shape content to appeal to them. This gave rise to the first generation of British comic book heroes such as Ally Sloper, iconoclastic characters who initiated a tradition of irreverence, paving the way for later staples such as *The Beano* and *The Dandy*.

This chapter will introduce readers to the early history of British comics. It will detail the context which made their publication possible by considering how a new audience was established via the migration from countryside to city (with the consequent rise of the industrial working class), the impact of new printing methods, and the effects of the Education Act. It will also show how comics drew on trends in popular print media such as book illustration and satirical cartoons which provide the cultural environment from which the comic could emerge. By examining the content of key titles, and the relationship they established with their readership, we will see how class identity was crucial to their appeal.

The audience for comics

The immediate popularity of early Victorian comics was dependent on the existence of a potential audience, already in place and waiting for them. This audience developed as a result of several economic and social changes which occurred during the nineteenth century. The first of these was the mass migration from the countryside to the quickly expanding cities. This movement had been growing steadily since the seventeenth century, provoked by increased efficiency in agricultural working practices, which meant that fewer manual workers were required on the land, and by expanding opportunities to import goods from abroad at competitive prices. The process accelerated sharply during the second half of the nineteenth century as 'the long and practically uninterrupted decline' in the number of men and women working in agriculture intensified (Thompson 1988: 25). In parallel to this, manufacturing industry was growing, particularly in the area of cloth production, providing the jobs which those same workers needed. These jobs were invariably to be found in the growing number of industrialised towns and cities; in 1801, there had been no towns with a population in excess of 100,000 outside of London but by 1901, there were nearly 40, and over a third of the population lived in them (Thompson 1988: 28). Smaller towns also grew in number so that by 1901, according to the census, three-quarters of the population were classified as living in urban areas, as opposed to just one-fifth a hundred years earlier (Mitchell 1988: 162). For historians like F.M.L Thompson, the move from agriculture to industry, and from the countryside to urban areas, was the single most significant change of the later 1800s and one which continued to shape the following century.

A consequence of this demographic shift was the creation of a new social grouping: the urban industrial working class. Formed through the coming together of complex economic and social forces during the nineteenth century, their character was also the product of their own response to those conditions. E.P. Thompson's landmark study, *The Making of the English Working Class*, although centred on the period 1780 to 1832, shows how the working class were instrumental in fashioning their own culture as distinct from other social classes. His account demonstrates how the working conditions of industrial factories and cotton mills, and the environs of the crowded tenements of the cities, brought into being a pattern of daily existence new to British society. Drawing influences from church, family structures, leisure activities, and increasing political activism, the working class began to form an outlook which differentiated them socially, forging a specific identity which then found expression through diverse cultural forms. Whilst E.P. Thompson is keen to celebrate this as a political tradition, and to recognise its formative role in shaping subsequent democratic dissent and reform, he also emphasises how a set of emergent working-class values could be found in everyday pursuits and even lowly amusements, a view significant to later popular publishing phenomenon such as the comic.

As Pamela Horn has shown, the growth of leisure activities was not only a major feature of the late Victorian period but one with a keenly 'class-conscious' character, demonstrated by a marked divide between what constituted pleasure for each class (Horn 1999: 9–12). James Chapman suggests that 'the late Victorian working classes had more disposable income and more time for recreation. A new leisure economy emerged to cater for this market' (Chapman 2011: 18). By the late 1800s, these pastimes could often be identified by the class origin of their audience, whether at the music hall or in professional sports like football and boxing, or at specific venues such as the public house. Even seaside resorts like Bournemouth or Blackpool could be classified according to the class of their clientele. Scott Banville has shown how the locations used as backdrops in *Ally Sloper's Half Holiday* form a map of these new venues linked by the class of its readership (Banville 2008). Crucially, for E.P. Thompson, these cultural developments are always political; writing of the working class he states that 'everything, from their schools to their shops, their chapels to their amusements, was turned into a battleground of class' (Thompson 1991: 914).

An additional factor in establishing the readership for comics among the working class was the reform of the educational system in England and Wales in the last 30 years of the nineteenth century. The movement towards reform, in all walks of life, was a transformative urge of the period. It can be found in political activism and the campaigns towards widening democracy, as evidenced by the sequence of Reform Acts throughout the century. It was also to be found in attempts to help the poor and to tackle the social ills that had arisen as a result of the population shift to the cities and towns. According to differing commentators, this was either the result of genuine philanthropy and Christian zeal by the middle class, or because of fear that a discontented working class might pose a real threat to the status quo. Typical of this is the Education Act of 1870, shaped by the Quaker W.E. Forster, which set out to ensure that everyone would be educated up to the age of 13. Even taking into account A.N. Wilson's suggestion that the working class were already actively educating themselves prior to this, the impact of the Act is hard to overestimate (Wilson 2002: 362). Its achievement was to remove the haphazard nature of educational progress for the masses and enshrine it in law and centralised practice.

By 1884, a potential audience for populist printed material such as comics was in place; this audience was the urban industrial working class. With sufficient finances to afford a cheap entertainment and enough spare time to enjoy it, and with standards of literacy to enable them to appreciate the form, a willing readership was available. This new entertainment now needed to find an appropriate style with which to appeal to the attitudes and values of its intended audience.

Popular printed entertainment in the Victorian era

In *British Comics: A Cultural History*, James Chapman suggests that the early history of comics is not so much a matter of various 'firsts' as an evolutionary process in which the conventions of the medium took shape across a number of publications (Chapman 2011: 17). To understand how we arrive at *Ally Sloper's Half Holiday* and its progeny, we need to map the formats which paved the way for its appearance. Printing in Britain can be dated back to 1476, with the first publications aimed at a wealthy, educated elite or burgeoning professions like the Law (Feather 2000: 8) but by the Victorian era it had developed into a genuinely mass media. During the eighteenth century, developments in transport had enabled the supply of printed material to spread away from London to the provinces, both fulfilling and provoking a huge increase in public demand. As John Feather puts it, 'it was from this cycle of development and change that there grew the great edifice of the Victorian publishing industry when the trade reached unprecedented, and perhaps still unequalled, heights of prosperity' (Feather 2000: 130). Increasing demand led to technical innovation which in turn led to faster, more efficient practices and lower prices. Techniques, which had remained largely unchanged for 400 years, underwent a process of mechanisation which completely altered the publishing industry. The initial focus for this change was the book, and more specifically the novel which reached new heights of popularity at this time. But the book was not the only medium transformed; other forms of printed material including newspapers, periodicals, and eventually the comic, were all radically affected by the ability to produce in greater numbers at a much reduced price.

The advent of the comic drew on a number of forms which had already developed in printed material earlier in the century. The first significant precedent is the use of illustrations in books. A variety of techniques had been pioneered in the early 1800s including wood engraving, etching, and various forms of lithography. This was extended further from the 1860s onwards by the introduction of photographic printing and the myriad types of illustration this made possible (Wakeman 1973). Photo-processes would eventually play a crucial role in the creation of comics in that they made it possible to produce multiple facsimiles of an artist's original drawing without recourse to an engraver. The resulting work ranged widely across natural history books with images of animals and plants, to travel guides, collections of maps, and expensive art volumes. Many of these were only available to the moneyed classes or else appealed to connoisseurs but the popular novels of the period sold across the social classes in great numbers. No one typifies this more than Charles Dickens whose stories, whether in complete book form or via periodical publication in instalments, attracted huge audiences and featured the finest illustrators of the day including George Cruikshank and Hablot Knight Browne, better known as 'Phiz'. Dickens himself recognised the importance of these illustrations in selling his books and took considerable care over his relationship with his illustrators (Muir 1971: 89–97). Many of the images produced are forerunners of the comic in their use of exaggeration and their ability to convey character and narrative with brevity for maximum impact. Cruikshank's engravings for *Oliver Twist* (1846) had such an impact that they affected the visual representation of characters like Fagin and The Artful Dodger through to the era of cinema; David Lean's 1948 version of the novel mimics the appearance of Cruikshank's illustrations.

An extension of this success was the use of illustration in children's literature, the popularity of which had an enormous influence on the style adopted in many late Victorian comics. The profitability of children's books had been established in the mid-eighteenth century but reached new levels in the next as the middle class began to educate their children at home using illustrated volumes. This educational approach was adopted by various religious groups who recognised the attraction of illustrations in drawing children to read bible stories (Feather 2000: 162–3), as well as for use in schools. In addition to educational books, fairytales, and folk stories were illustrated for children, as were new works of fiction for this expanding market such as Lewis Carroll's *Alice's Adventures in Wonderland* (1865) and *Through the Looking Glass* (1872). The illustrations for the first editions of Carroll's works were by John Tenniel whose distinctive images indicate a tension at the heart of Victorian's children's book illustration, which was to be found in later comics. The intention of these books was often to instruct children in the moral standards of the day. However, to attract the intended readership anarchic humour was to the fore. These twin strands could be contradictory, as is apparent in Carroll's books which satirise the 'improving' nature of children's literature of the period. It is also evident in Tenniel's illustrations which are wildly surreal, a quality taken up later in the fairytale artwork of Arthur Rackham.

Another rich source of inspiration, and sometimes future personnel, for the nascent comic book was the newspaper cartoon. As Keith Williams shows, the newspaper publishers and editors of the previous century initially frowned upon illustration:

> Papers were designed to appeal to a readership that enjoyed the elite status of literacy in a largely illiterate population. Those who had mastered the skill of reading would not wish to share their understanding with those who could only look at the pictures.
>
> (Williams 1977: 79)

By the time comics arrived, such questions of literacy had become redundant but the notion that looking at pictures was for the working class still held sway. Newspapers had become affordable to the working masses with the reduction of their cover price following the abolition of stamp duty on newspapers in 1855 and then the removal of the paper excise duty in

1861. The first newspaper illustrations were political cartoons but by the mid-1800s, weekly papers such as *Illustrated London Life* and *The Graphic* had built a considerable following with their use of images. Artists were employed to illustrate topical scenes to add excitement to reportage, an approach which gained greater success with the advent of photography. These 'actuality' illustrations were to transfer over to a number of the early comics and had their continuation in later publications like *Eagle* (1950–69 and 1982–94) with its memorable 'cutaway' panels showing the inner workings of motorcar engines or fighter planes. The satirical cartoon became one of the most powerful forms of the Victorian era, establishing a British tradition upheld to this day by artists like Gerald Scarfe at *The Sunday Times*. Drawing on antecedents such as William Hogarth's popular moralistic engravings, the Victorian cartoon attacked the follies of successive governments of differing ideologies and ridiculed failings in foreign policy such as the military mishaps of the Crimean War (1853–6). The breadth of the satirical attack is evidenced by the fact that cartoonists could both mock the British Empire itself or just as equally ridicule the inability of governments and civil servants to run it properly (Douglas 1993).

Such work became a feature of the growing raft of illustrated magazines which became successful in the late nineteenth century. Typical of the satirical tradition was *Punch*, founded in 1841 by the writer Henry Mayhew and the engraver Ebenezer Landells; it survived until 2002. For readers of *Punch*, visual illustration was every bit as important as text. As well as being directly political, the target for many cartoons was the foibles of polite Victorian society, from modes of dress and speech to domestic family life. *Punch*, along with other contemporary satirical journals, offered two dominant modes of illustration (Perry and Aldridge 1989: 32). The first was the single-panel cartoon, often with a humorous caption underneath and occasionally with dialogue carried in a speech balloon or bubble; the drawing style was detailed with a tendency towards exaggeration. The second form was the strip cartoon where a short sequence of images, perhaps just two or three, are shown together to convey a small narrative or a more complex joke. All of these elements were to transfer to the comic. It is around this time that the term 'comic' probably came into use as publications which put an emphasis on humour, with the use of illustrations, became known popularly as 'the comicals' or 'comics' (Sabin 1996: 12).

However, the readership of *Punch* was the middle-class intelligentsia rather than the masses. It was vulgar and sensational publications which reached a wider audience, such as *The Illustrated Police News* which began in 1864. This weekly paper focused on grisly crimes, illustrated with gaudy panels showing violent murders or the equally unpleasant retribution of the courts in the form of public hangings. Among its most notorious editions were its coverage of the Whitechapel murders of 1888, as well as its xenophobic depiction of foreign 'aliens' intended to play on the fears of its urban working-class readership. A characteristic page from the Jack the

Ripper coverage entitled 'Dr Forbes Winslow conjures up the secret action of Jack the Ripper' unfolds as a comic strip with panels reading left to right down the page as the narrative unfolds. Such publications quickly acquired the title of 'penny dreadfuls' in recognition of their commercial exploitation of the public's interest in sensation. The true penny dreadfuls, such as *Black Bess* (1863) and *Sweeney Todd, the Demon Barber of Fleet Street* (1878), featured fictional stories which were presented mainly as text but with vivid illustrative panels. Their direct appeal to a predominantly working-class readership, as well as the bluntly direct mode of address used, became typical of the later Victorian comics and influenced the development of the tabloid press. In direct contrast, titles like *Boys Own Magazine* (1855) and *The Boy's Own Paper* (1879), the latter published by the Religious Tract Society, used a similar format to present material intended to uphold conservative moral values and extol the virtues of patriotism and the Empire. Although largely middle class in terms of their readership, some boy's papers such as *Boys of England* (1866) set out to win a more working-class audience.

By the late nineteenth century, various forms of printed entertainment were available to the public, propelled by improvements in technology which made it possible to produce books, papers, and magazines in large quantities and sell them at modest prices. The skills of talented writers and illustrators had been honed and formats had been established with a proven track record of popular appeal, from actuality to satire. With a waiting mass audience, conditions were set for the appearance of the comic.

Ally Sloper and the birth of British comics

Accounts of the development of British comics invariably suggest that the first regular publication to be recognisable as a comic in the contemporary sense is *Ally Sloper's Half Holiday* (Perry and Aldridge 1989: 47). Even if this is sometimes disputed, for Denis Gifford Ally himself 'qualifies as the first true British comic strip hero' (Gifford 1976: 9). The comic's first issue appeared on 3 May 1884 and was published in London by the Dalziel brothers; the paper was to continue until 1923. Published weekly on Saturdays and priced at one penny, it owed much to the ground laid out by *Punch* but shifted its focus decisively on to visual illustration, being liberally decorated from cover to cover with images, leaving less room for blocks of printed text. The emphasis was on single panel, standalone gags but frequently a running joke would be elaborated in variations over a number of individual panels. From the beginning there were also extended sequences of more tightly related images conveying a frequently convoluted narrative which are closer to the comic strip format we would recognise today, although usually shorter. Broad humour dominated in a style more physical and crude than that normally seen in *Punch* or its rivals but the influence of the illustrated magazines was still strong, and the paper featured a good number of educational and traditional journalistic pieces, albeit in visual form such as travelogues. There was also still a large amount of text by the standard expected in today's comics.

Ally himself had first appeared as a character in one of *Punch*'s rival satirical titles, *Judy*, in 1867. He was created by Charles Ross and drawn by his wife, Isabelle Émilie de Tessier, under her pen-name, Marie Duval; she was probably the first female comics artist. From the start, his characterisation centred on what David Kunzle describes as 'lower-class alienation from and rejection of the traditional concept of useful, humble work', a concept which he suggests was perpetrated by bourgeois culture to reinforce a reassuring image of the dutiful working classes (Kunzle 1985: 40). Ally then became the first cartoon character to acquire his own weekly paper in the form of the *Half Holiday* where he was initially drawn by the American-born W.G. Baxter, who was then followed in 1888 by W. Fletcher Thomas, after the death of Baxter, and subsequently by C.H. Chapman. The comic's full subtitle described it as 'Being a selection, side-splitting, sentimental, and serious, for the benefit of old boys, young boys, odd boys generally, and even girls'. In essence the paper was an amalgamation of many of the most popular elements which had appeared in illustrated periodicals over the preceding 50 years but with an emphasis on visual humour, rather as if an editor today took over *Private Eye* and then removed all of its pages of tightly displayed text leaving only the cartoons and comic strips.

As Roger Sabin has shown, the comic quickly became a publishing phenomenon, an achievement that was considerably aided by skilful marketing (Sabin 1996: 18–19). It proved to be particularly popular reading on railway journeys and was heavily promoted through kiosks on railway stations, becoming part of a group of titles known as 'railway literature'. The publishers even offered accident cover for anyone found in possession of the paper when such an unfortunate incident had occurred. Merchandising in the form of mugs, dolls, watches, postcards, and posters were highly successful and helped to promote sales, as did various special offers made to regular readers in the pages of the comic, creating an economic model for future similar publications. Ally's success even crossed over into other media such as films and the stage (Sabin 2009). The paper was soon able to claim, with justification, that it was the biggest selling penny paper in the world.

A major factor in the success of *Ally Sloper's Half Holiday* was its direct appeal to a specific audience. Unlike *Punch*, its readers were not the middle-class intelligentsia but, as Perry and Aldridge put it, the 'young men, clerks and counterhands – the new generation of white-collar workers' (Perry and Aldridge 1989: 54). This readership was also markedly younger than that of *Punch*, although the *Half Holiday* was still aimed at adults rather than children. Peter Bailey's meticulous research into the readership for *Ally Sloper* indicates evidence for its popularity among some middle-class readers, and even occasionally with the social elite of the period, but tends to support Perry and Aldridge's view that the overwhelming majority of the readership was working class (Bailey 1983: 9–10). The crucial element in this appeal seems to have been audience identification. For Perry and Aldridge, Ally was a 'seedy proletarian loafer' whose anarchic humour 'and contempt for stuffed shirts and pomposity were directed towards the new literate masses emerging as the first effects of the 1870 Education Act' (Perry and Aldridge 1989: 47). For James Chapman, he embodies every Victorian middle-class prejudice about the working classes, as he is 'feckless, work-shy, a scrounger and a rogue' (Chapman 2011: 20). His very name, like many of the characters in the novels of Charles Dickens, was an indication of his personality in that he was likely to be found sloping off work or down a back alley to avoid the rent collector. Yet his behaviour isn't condemned in the pages of his journal; it is celebrated.

Scrutinising the paper's early issues where he is invariably to be found hatching a scheme to get rich quick, or at least avoid hard work, easily identifies the class-based nature of his appeal. On the front page of the issue dated 6 May 1888 we see 'Ally at the Royal Academy'. In characteristic fashion he has managed to find a painting of three female nudes and has sat down with a picnic basket and binoculars to fully enjoy the view. Unfortunately for him, his stern-faced wife, with baby carriage in one hand, is seen apprehending him by the ear. Such was his folk hero status that by issue 344, dated 29 November 1890, the police who have descended to arrest him are then unable to hold back the tears at the ordeal of having to apprehend this class hero. His anti-establishment tendencies could sometimes be directly satirical, and commented on topical events such as the cover story for 9 August 1884 which refers to the Tichborne lawsuit with Ally, suggesting that the claimant should be released so that he could enjoy the pleasures of the seaside season. Ally's male proletarian characteristics – evading both employment and domesticity so as to head for the nearest pub, bookies, or any kind of sporting activity (strictly as an observer), motivated by laziness and the pursuit of easy pleasure – established a comic model which recurred repeatedly in the British low comedy tradition afterwards. It can be seen a century later still at large in the form of Sid James of the Carry On team, in the failed schemes of Del Boy in the television sitcom *Only Fools and Horses*, or in the *Daily Mirror*'s much-loved cartoon strip *Andy Cap*.

Ally, and many of the cartoon characters and comics which followed in his wake, quickly attracted criticism. The nature of the attacks, which came invariably from middle-class commentators, tends to confirm the appeal of these comics to a working-class readership. They were criticised as being a bad influence on the proletarian masses, setting back the cause of literacy with their emphasis on visual storytelling and pandering to the ignorance and prejudices of their readers. As Roger Sabin has suggested (Sabin 1996: 19), the language used by the critics tends to reveal their own class-based position as they frequently referred to the comics and their readers as 'vulgar'. More recent examples in the tradition like the contemporary adult comic *Viz* were to deliberately play upon this reaction and maximise the potential for causing offence. Perhaps the criticism accounts for a gradual alteration in Ally's character as, in his later incarnations, 'he smartened up his wardrobe and, shaking off his criminal leanings and working-class roots, became "the Friend of Man" or a man of the people, and something of a gent' (Gravett and Stanbury 2006: 18). As a result, later strips show him in the company of the upper classes, and even the Royal family, effectively robbing him of his radically subversive impact.

Comic Cuts, Chips, and the beginnings of a tradition

The success of *Ally Sloper's Half Holiday* opened the door for publishers to exploit the same format and reach the same potential audience. The first to do this consistently was the publisher Alfred Harmsworth, owner of what became the Amalgamated Press, who was especially keen to counterbalance the popularity of the penny dreadfuls, publications that he strenuously disliked (Perry and Aldridge 1989: 47). He began with *Comic Cuts* in 1890, the first publication to use the term 'comic' in its title; the phrase 'comic cuts' was already in circulation having been used to denote the humorous images, taken originally from wood cuts or engravings cut into metal, which interrupted the prose text of more traditional papers and magazines. In Harmsworth's publications, the humorous images more or less took over. Initially, *Comic Cuts* was designed as a crude means to direct potential readers towards another of Harmsworth's publications, *Answers*, which was currently being outsold by its chief rival in the shape of George Newne's *Tit-Bits*. However, the plan backfired in the best way possible: its first edition sold over 100,000 copies and it grew within weeks to a regular run of 300,000, outselling *Answers* by more than 100,000 (Perry and Aldridge 1989: 48). Illustrated *Chips* quickly followed *Comic Cuts* later the same year. Both titles were to run for more than 60 years, and in their prime the two together had a circulation of more than half a million. An important element in the success of Harmsworth's comics was that he cut the cover price in half, with titles going on sale for a mere halfpenny. The strategy was based on a prediction of consequently high sales and by the increasing efficiency of printing techniques which would reduce production costs. It was a gamble that paid off.

Harmsworth's publications imitated the style established by *Ally Sloper's Half Holiday*, with a mix of single-panel gags, shorts strips, and prose stories. Most significantly, they drew on the idea of creating and sustaining comic characters with which the working-class readership could identify, and who they would want to follow week after week. The most popular of his comics stars were the duo Weary Willie and Tired Tim who appeared in *Chips*. Created by Tom Browne, who became one of the most important and prolific of the early comics artists, it featured two likeable tramps whose attempts to avoid gainful employment and live a life of relative ease bore considerable resemblance to Ally Sloper and offered much the same appeal. Slapstick is again to the fore as our heroes hatch various ambitious plans which invariably fail. For issue 298 dated 16 May 1896, they try to obtain an easy ride home by tying a horse to the front of a dinghy and persuading it to tow them back along the river, only for the animal to become distracted and tip them into the drink. In issue 528 from 13 October 1900, they abscond with a steam train in order to escape the attentions of the police, only to inexplicably end up at the North Pole where they then sell the train to a group of explorers for a handsome margin. Despite many mishaps they were, in the words of comics historians Gravett and Stanbury, 'far from being pitiable failures, tramps were seen by many as dignified dropouts in an era when industrialisation rewarded millions with poor pay, working conditions and health' (Gravett and Stanbury 2006: 19). Like Ally, they rapidly assumed the status of class heroes providing a template for Charlie Chaplin's even more popular tramp.

Comic Cuts similarly drew on its appeal to a working-class readership, billing itself with considerable pride as 'the poor man's *Punch*'. Promising 'one hundred laughs for a halfpenny', its humour was largely broad slapstick drawing on the approach of comedians from the music hall. The cover of its first edition from 17 May 1890 features half a dozen short strips and single panels including a strong man whose ability to lift a 1,000lb weight is revealed as a trick when a dog walks off with the air-ball in question, and another strip showing a bossy husband falling from a step ladder while hanging a picture. However, a class emphasis is apparent in several jokes and drawings which mock the middle classes, as in the picture of 'the joys of the suburbanite' where a respectable couple are seen dashing for the morning train, or another showing Tingford, a social climber who has moved to Chingford – 'an awfully jolly place' – only to have his daily constitutional wrecked by its relentlessly appalling weather. It soon fell into the pattern established by *Chips* and began to feature regular characters in longer strips that covered a whole page. Typical of this are the anarchic Chokee Bill and Area Sneaker, another pair of scheming working-class layabouts obviously modelled on Weary Willie and Tired Tim.

In the wake of *Ally Sloper's Half Holiday*, *Comic Cuts* and *Chips* came a battalion of imitators, often of inferior quality, to which Harmsworth responded with even more titles of his own. The newsagent's shelves bulged with *The Funny Wonder* (another title from Amalgamated Press), *Larks!* (from the Dalziels), *Comic Life*, *Jolly Bits* and *Funny Cuts*. To keep costs down to a halfpenny, these publications used exceedingly poor paper and ink, establishing a reputation for parsimony which stayed with British comics for many years after and which was initially seized upon, and partially parodied, by *Viz* before it gave in to the contemporary market place and turned glossy. The link to other popular working-class entertainments was also established with *Dan Leno's Comic Journal* (1898) featuring a cartoon version of the highly successful music hall comedian. By 1900, the basics of a regular format had been established across these titles. On the front cover was a comic strip which covered the entire page, typically in six panels, which was then followed by a few pages of prose stories, often tales of daring adventure, with illustrations. A double-page centre spread featured short strips and single cartoon gags, then there were more stories and then a couple of popular strips on the back page. By the new century, some titles had already begun to introduce colour such as James Henderson's *Comic Life* and Amalgamated Press's *Puck* and *Chuckles*, as well as to differentiate their readership by producing titles specifically aimed at younger children such as *Rainbow*; these were to establish a huge following in the 1920s. Although *Ally Sloper's Half Holiday* was clearly aimed at adults, titles such as *Chips* and *Comic Cuts* had moved the age demographic lower, so that the readership gradually shifted towards children. In doing this, a style was established that would lead in due course to the golden age of British comics, with a readership largely consisting of children, which reached its height with the mass popularity of *The Dandy*, *The Beano*, *The Topper*, and many other titles (most of them published by D.C. Thomson) from the 1930s onwards. These titles often retained the physical humour of their predecessors, as well as their anarchic spirit and dislike of pomposity albeit with their class affiliations softened to broaden their appeal.

Conclusion: comics and class

For historian Jose Harris, the last quarter of the nineteenth century is the period 'in which the tentacles of class became all embracing, in which all other social and cultural attributes became reducible to class categories' (Harris 1993: 6). Changes in the social and economic structures of Britain produced a new urban working class who were increasingly literate and educated, as well as being discontent with their lot. This may have found its higher flown manifestation in the rise of the trade unions and the creation of the Labour Party, or in the movement towards self-improvement, which the working classes fostered, but it also led to the development of a popular culture, and leisure pursuits which were distinctive in character. These provided a respite from the grind of hard work, poor wages, and bleak living conditions but crucially they also gave voice to those feelings of resentment, frequently through humour in the form of mockery aimed at those in positions of authority. It could also enable the working class to play out escapist fantasies of a life of idleness and pleasure which was normally denied them. All of the elements found their focus in the cheap journals and papers now made easily available to them by advances in technology. If the comics rapidly altered their readership towards children as the new century dawned, perhaps recognising something of the same anti-authoritarian and anarchic tendencies in the young, then for a period at the end of the nineteenth century, they spoke directly to an adult working-class audience who found both enjoyment and a form of rebellion in their populist pages.

References

Bailey, P. (1983) Ally Sloper's Half Holiday: comic art in the 1880s. *History Workshop*. 16. 4–31.

Banville, S. (2008) Ally Sloper's Half Holiday: the geography of class in late-Victorian Britain. *Victorian Periodicals Review*. 41 (2). 150–173.

Chapman, J. (2011) *British Comics: A Cultural History*. London: Reaktion.

Douglas, R. (1993) *'Great Nations Still Enchained': The Cartoonists' Vision of Empire, 1848–1914*. London and New York: Routledge.

Feather, J. (2000) *A History of British Publishing*. London and New York: Routledge.

Gifford, D. (1976) *Victorian Comics*. London: Allen and Unwin.

Gravett, P. and Stanbury, P. (2006) *Great British Comics: Celebrating a Century of Ripping Yarns and Wizard Wheezes*. London: Aurum.

Harris, J. (1993) *Private Lives, Public Spirit: Britain 1870–1914*. London: Penguin.

Horn, P. (1999) *Pleasures and Pastimes in Victorian Britain*. Stroud: Sutton.

Kunzle, D. (1985) The first Ally Sloper: the earliest popular cartoon character as a satire on the Victorian work ethic. *Oxford Art Journal*. 8 (1). 40–48.

Mitchell, S. (ed.) (1988) *Victorian Britain: An Encyclopedia*. London: St. James Press.

Muir, P. (1971) *Victorian Illustrated Books*. London: Portman Books.

Perry, G. and Aldridge, A. (1989) *The Penguin History of Comics*. London: Penguin.

Sabin, R. (1996) *Comics, Comix and Graphic Novels*. London: Phaidon.

Sabin, R. (2009) Ally Sloper: The First Comics Superstar? In Heer, J. and Worcester, K. (eds.). *A Comics Studies Reader*. Jackson, MS: University Press of Mississippi.

Thompson, E.P. (1991) *The Making of the English Working Class*. London: Penguin.

Thompson, F.M.L. (1988) *The Rise of Respectable Society: A Social History of Britain, 1830–1900*. London: Fontana.

Wakeman, G. (1973) *Victorian Book Illustration*. Detroit, MI: Gale.

Williams, K. (1977) *The English Newspaper: An Illustrated History to 1900*. London: Springwood Books.

Wilson, A.N. (2002) *The Victorians*. London: Arrow.

11 The spectacle of speech

Victorian popular lectures and mass print culture

Anne-Julia Zwierlein

Public speech events were omnipresent in all cultural fields during the Victorian period, in politics, higher education and rational recreation. Martin Hewitt reminds us of the ubiquity and dazzling variety of platform performances in nineteenth-century Britain:

> At mid-century, even a relatively modest town would have been the venue of an average of more than one public meeting a night; in the larger towns and cities by the 1860s and 1870s there might easily have been half a dozen daily. Most congregations heard two sermons each Sunday, and quite frequently a religious lecture during a week night, and a bewildering array of associations, from parochial young men's societies to political clubs, organized popular lectures, and debates.
>
> (Hewitt 2002: 2–3)

Traces of these ephemeral rhetorical performances are accessible through the rhetoric of texts, through journalistic coverage in newspapers and periodicals, lecture notes, epistolary or (auto)biographical reports, and mediated versions of public speech in fiction writings. The sheer abundance of Victorian platform production is, as Hewitt concedes, "overwhelming": "it was an element of public life that defied quantification" (2). Sermons, courtroom speeches, author readings, political addresses and public lectures were avidly consumed in both metropolitan and provincial sites. Depending on venue and institutional setting, the sociological constitution of audiences extended from the working to the upper classes. Outside the context of working-class political agitation (from the Chartist movement of the 1830s and 1840s to the extension of suffrage in the 1867 and 1884 Reform Bills), audiences of public speech events from mid-century onwards were dominated by the aspirational labouring population and by the middle classes eagerly bent on self-improvement – including, increasingly, women (whose own fight for suffrage was on the horizon).

Taking into account that the culture of public speech was comprised of oral performances as well as their journalistic coverage, overall audience numbers could be huge: easily thousands of listeners indoors as well as outdoors in the case of attractive or controversial topics (see ibid.: 2, and Meisel 2001: 262), multiplied by the hundreds of thousands of readers who followed speech events in the newspapers. Indeed, the circulation numbers of the larger dailies could reach between half and one million (see King 2004: 84, and Ellegard 1971), and we know that each paper copy was shared statistically by a sizeable number of people – partly owing to the fact that articles were read aloud in the family circle or workplace (see King 2004: 120, and Rose 2001: 87–88). Thus oral speech travelled through the medium of print in order to resurface in other oral performances. The Victorian social practice of rhetoric can therefore be envisaged as a mass leisure culture that held its own against, and was inextricably tied up with, the culture of mass print. Of course the expansion of literacy and the proliferation of print products in the second half of the nineteenth century was a decisive cultural shift. Yet when Andrew King and John Plunkett, following Walter Ong, claim that "print came to dominate [nineteenth-century] British society to such an extent that oral forms of culture lost much of their status" (King and Plunkett 2005, 1; see Ong 1982/2005), this underestimates the continuing importance throughout the century of the oratorical and the auditory, in both elite and popular culture. This chapter conceives of nineteenth-century culture not as "a society's 'Being-in-Print'" (King and Plunkett 2005, 1), but as characterized by a mutual influence, a feedback loop between the oral and the written. Indeed, Robert Ellison insists on the "residual orality" of Victorian discourse (1998, 132), and Joseph Meisel confirms that "the practices of public speaking as they developed in nineteenth-century Britain were a *new* formulation strongly connected to a host of other modernizing trends in the period" (2001: 5). Increasingly, the public speaker's and lecturer's personality and platform presence became part of the new spectacle of orality, and the late-Victorian sensationalist "New Journalism" displayed a similar fascination with speakers' appearance and modes of delivery (see Hewitt 2012: 86–87, and Wiener 1988). These new kinds of celebrity attached not only "to the notable orators of the period", the focal point of Meisel's study on public speech (2001: 278), but they inhered in the culture of orality in general. Although, according to Joseph McCabe, by 1870 the "high point of the culture of public speech" had passed (1908: 285), this was the point when oratory had completed the transition from "great speeches" by "great men" (Meisel 2001: 9) to a more comprehensive, popular culture of public speeches and lecturing.

From the vast variety of Victorian oral performances, this chapter singles out the popular lecture for particular attention – more specifically, the popular science lecture which originated with the itinerant natural philosophers of the eighteenth century and was thus the oldest form of this oral (and visual) genre. Travelling through the Romantic literature and arts institutes and the early-Victorian Mechanics' Institutes, it was finally integrated into late-Victorian middle class rational entertainment. Indeed, "the popular lecture participated in some of the most pervasive characteristics of Victorian science culture, its exhibitionary impulses, its museological textures and its luxuriance of display" (Hewitt 2012: 80; on the "showiness" of science culture, see also Fyfe and Lightman 2007; Lightman 2007; O'Connor 2007). Here, too, audience numbers could be enormous – as Hewitt confirms:

> even apparently dry academic subjects could draw huge crowds, such as the 3,700 that attended John Tyndall's lecture on "Crystalline and Molecular Forces" in the Manchester Penny Science series in the 1870s, and it was claimed that 2,000–3,000 was a common figure for the science lectures organized around Britain in the 1880s and 1890s by the Gilchrist Trust, with one London course reaching a peak of over 5,000. (2002: 4)

This chapter explores how ephemeral lecture performances were modelled in newspaper reports and popular fiction, mainly between 1860 and 1900. Astronomical lectures – the majority of the examples in this chapter – offered a fascinating mixture of orality and visuality owing to the use of experiments and illustrations. Conjoint questions are how lectures created authority and legitimacy for the speaker, and how moments of lecturer–audience (and audience–audience) interaction were involved with the problem of attention, a vital topic at the root of "modern" psychological constructions of subjectivity. Jonathan Crary maintains that it was precisely the new bourgeois emphasis on disciplined attention that made failures of attention visible: "modern distraction [...] was an *effect* [...] of the many attempts to produce attentiveness in human subjects" (2001: 49). This chapter suggests that non-fictional and fictional accounts of popular lecture performances can afford us glimpses of an alternative history, one in which the voice of the lecturing authority becomes a backdrop for the voices of the audience who are directing their attention differently or elsewhere. These instances of mediated orality and audience responses can also help us rectify earlier "projection[s] of an undifferentiated public that would be better understood in terms of competing readerships or counterpublics", as Matthew Rubery phrases it in his revisionary analysis of the late-Victorian conservative discourse about "press decline" (2009: 10–11). Popular lectures responded to, and helped to condition, wider changes in British society and culture, and textualized oral performances articulated and negotiated pressing late-Victorian social and political concerns about the nature of authority, class, and gender.

The popular science lecture: radical or apolitical?

Popular science lectures had been "a feature of the distinct industrial culture of provincial towns in eighteenth-century Britain", starting with the itinerant natural philosophers who "instructed employers and artisans alike on matters directly pertinent to their lives (especially mechanics, light, and heat)". Because these lecturers were private entrepreneurs, their performances had to be both educational and entertaining: they "created a culture of 'shocks and sparks' that made the science lecture sensational and drew audiences well outside the industrial areas" (Leaney 2008: 160–61; see also Schafer 1983; Inkster 1981; Inkster and Morrell 1983). Far into the 1820s, public meetings and lectures were perceived as potential threats to the "social order" (see Hewitt 2002, 6; 8), and the same suspicions were provoked by the Mechanics' Institutes which were founded throughout Britain during the first half of the nineteenth century (on their rapid expansion and increase in membership, see Altick 1957: 190). Channelling the functions of the itinerant lecturers by conveying scientific knowledge to workingmen through classes, lectures, and institute libraries, they advertised science as "useful (and apolitical) knowledge" (Hewitt 2012: 82). This was surprising given the earlier, Romantic association of science with radicalism – and indeed, notwithstanding their conservative agenda, the institutes were regarded askance by the Tory party and the Church of England (on the conservatism of the institutes, but also on Tory suspicions of radicalism, see Altick 1957: 188–94; Leaney 2008: 168; Altick 1978: 368). In fact, they never achieved any kind of mass appeal, as the social-disciplinary aspect of their programmes seems to have been all too obvious. Richard Altick reminds us of a mechanic's pertinent assessment of the institutes' middle-class ideology: "You must remember we have masters all day long, and we don't want 'em at night" (1957: 192–93, quoting from Henry Solly, *These Eighty Years*, 1893).

During the last third of the century, the Mechanics' Institutes were consequently taken over by a predominantly middle-class audience, and they were often transformed officially into "Literary and Scientific Institutions": a famous and representative example is the Birkbeck Mechanics' Institute in London which was rechristened the Birkbeck Literary and Scientific Institution in 1869. Reinterpreting lectures as mass leisure culture, they retained (some) of the earlier educational agenda, but also, with varying degrees of explicitness, the veto against political topics. While evening classes increasingly addressed daytime professionals, the afternoon lectures and courses soon became the preserve of a women audience. Kate Newey sees this as part of a process of democratization: "working-class men had civic visibility; meanwhile, women of all classes were increasingly visible and independent in public. [...] popular entertainments were constructed especially to meet the demands and desires of these various constituencies" (2010: 157). This broadening of audiences also caused a shift away from science lectures, as Altick observes: "in conspicuous contrast to the formal scientific discourses envisaged by the founders, [they] were concerned less with 'useful arts' than with history, travel, biography, music, and literature" (1957: 201). While there were always some "star" lecturers like Michael Faraday, who gave his famous electricity lectures to the Royal Institution during the 1820s, and later John Tyndall, T.H. Huxley, or Robert Ball with his "popular expositions of astronomy in the 1880s and 1890s [...] to audiences both specialized and general", from the mid-Victorian period onwards science lecturers formed "only a small fraction of the whole" (Hewitt 2012: 81–82).

However, the miscellaneous Victorian popular lecture culture managed to integrate science into new formats: besides the more sober lecturing institutions, there was a huge array of multi-media "shows" catering to the metropolitan masses. The London Royal Polytechnic Institute, established in 1838 with the aim to encourage invention and technology, soon came to specialize in sensational illustrated lectures, from the famous "Pepper's Ghost" shows to "spectacular displays of scientific wonders, including optical illusions demonstrated by the Polytechnic's numerous magic lanterns" (Groth 2013: 103; on John Henry Pepper's spectacles, see also Flanders 2006: 271). In the Great Hall, other attractions were added to the lecture theatre, such as "a diver submerged in a diving bell", an oxyhydrogen microscope and other scientific and optical instruments, "large electrical machines in action", and "experiments conducted in a purpose-built laboratory" (ibid.). As a hybrid between exhibition and lecture institution, the Polytechnic "compet[ed] for market share with [...] the Egyptian Hall, Madame Tussaud's, and Wyld's Model of the Globe" (ibid.; on the Egyptian Hall see Featherstone 2002). Similarly, popular venues in this diversified mediascape were the Coliseum in Regent's Park and Robert Barker's panorama building in Leicester Square. It is important to remember, however, that science lectures, from the itinerant days onwards, had always been reliant on a mixture of education and entertainment. Altick claims that audiences at the Mechanics' Institutes, too, "were bent on achieving the illusion of being instructed when in reality they sought, and obtained, mere amusement" (1957: 201). Yet his argument that lecture venues only became successful once "the utilitarian motive was diluted and, eventually, expunged" (1957: 188), as well as Judith Flanders' comparable observation that the Polytechnic's "educational and scientific demonstrations and lectures had been diluted and were in practice indistinguishable from the entertainments of the town" (2006: 271), should provoke some suspicions about their use of the term "dilution", with its implicit assumption of a "pure", original form of science lecturing. We will, therefore, now turn to the technicalities of the Victorian lecture circuit, inquiring more specifically into how they conditioned the authority, legitimacy, and "popularization" of science.

Voice and authority: the popular (science) lecture and "popularization"

In the course of the nineteenth-century processes of urbanization and democratization, new norms of authoriality for the lecturer persona emerged. With the gradual extension of suffrage and the foundation of the British Association for the Advancement of Science, larger segments of the population had become enabled "to reach their own informed opinions if not to superpose these upon [the evidence of politicians and men of science]" (Sleigh 2005, 19, with reference to Morrell and Thackray 1981). Charlotte Sleigh compares physicist Faraday's tightly orchestrated lectures, his masterful control "over what was seen by his audience and what was not", with the strategies of popular sensation fiction, which was both authoritative *and* in collusion with its readership (ibid.: 22; on Faraday's lectures see Morus 1992). She also comments on "scientists' moves to subsume moral authority at the end of the nineteenth century" (ibid.: 25). However, authority had to be established through the lecture performance, and it remained a challenge to stage the visual (and auditory) effects of science lectures outside a small number of dedicated spaces like the Royal Institution or the Royal Polytechnic (see Hewitt 2012: 82). Experiments would miscarry, equipment would be dysfunctional; even the magic lantern, often used for illustrations from mid-century onwards, was accident-prone (see ibid.: 84). Care also had to be taken to dissociate the science lecture from fairground magic, as Barbara Stafford has shown (1994); and Hewitt confirms: "Too sensational, and the experimental lecture risked association with the disreputable appropriations of 'Electro-Biology' or mesmerism. Insufficiently glittering, and science lectures were vulnerable to dismissal as rather arid and mechanical" (2012: 83). In order to become moral as well as scientific authorities, irrespective of venue and equipment, lecturers, therefore, had to embody their message. Many lecturers reminisced about the physical effort that was needed to sustain a performance: "'An hour's talk' from Peter Mackenzie 'would leave him bathed in perspiration even to his outer garments'" (ibid.: 92, quoting Dawson 1896, 270; on the strain of lecturing, see also Meisel 2001: 263–64). Huxley flirted with his own exhaustion in his famous lecture on cell dissolution, "On the Physical Basis of Life" (1868), where he imagined himself, later at dinner, replacing the bodily matter he had 'used up' during the lecture by "recourse to the substance commonly called mutton" (1868/1968:147). Successful lecturers emphasized the physicality of their performance in various ways, pacing the platform, raising their voices, and gesticulating. Tyndall, famously, integrated a one-time near-accident as a regular feature: knocking over, and quickly catching, a flask (see Lightman 2007: 177). Many lecturers also complained about insensitive chairmen "who will remain on the platform", thus spoiling the dramatic impact of their performance (Hewitt 2012: 89). Lecturers' bodily investments served the purpose of establishing their sincerity and authenticity, and they returned the performance, despite the charts, diagrams, and lanterns, time and time again to its centre: the lecturer, and the spoken word (see also Plunkett 2015).

Behind these strategies of verbal and embodied authority was the implicit acknowledgement that the message might fail to reach the audience. Roland Barthes, in an 1971 essay translated as "Writers, Intellectuals, Teachers" and cited by Caroline Rose in conjunction with her analysis of Charles Kingsley's public speaking, took the Foucauldian perspective that the "speaker on the lecture platform resembles [...] a 'policeman' because [...] any attempt to disturb the clarity (and legality) of the message inevitably involves speaking 'badly'" (Barthes 1971: 192, 194). Language, he claimed, "is always a matter of force, to speak is to exercise a will for power". Yet Barthes did concede a degree of independence to the audience: "when the teacher speaks [...], the Other is always there, *puncturing* his discourse" (Barthes 1971: 194; quoted in Rose 2002: 136). Rose confirms this observation with a view to Kingsley's gruelling lecture tours and their physical and emotional toll, as reported by his wife:

> he seldom returned from speech or lecture without showing that so much life had actually gone out of him – not only from the strain of brain and heart, but from the painful sense of antagonism which his startling mode of stating things called out in his hearers, and of which he was keenly conscious at the time.
>
> (quoted in Rose 2002: 147)

The audience is only seemingly silent, and the speaker's body and words are "exposed" on the platform. Kingsley's stage fright, and his rhetorical feminization of his own activity as science popularizer (comparing it to that of [female] camp followers picking up leftovers from an army), are indicators of how keenly he sensed the precariousness of lecturing authority (see ibid.: 144–45). Similar strategies – and anxieties – are evident in John Ruskin's *The Ethics of the Dust: Ten Lectures to Little Housewives on the Elements of Crystallization* (1865), which pretends to record lectures that had never been delivered: the text alternatingly establishes distance and proximity between Ruskin's author persona, staged as an exhausted "Old Lecturer (of incalculable age)", and a (ventriloquized) lively group of schoolgirls (Ruskin 1865: 18:207; see Zwierlein 2009: 91–128).

Greg Myers assumes a clear hierarchy of communication for didactic science "dialogues" (and, by implication, educational lectures) – "Because they were usually written for women or children they raise problems of how scientific culture is presented to those presumed to exist outside it" (1989: 172; examples of such pseudo-dialogues, with their staged orality, are Joyce 1807 or Kingsley 1869). Yet fictionalized lectures often tell a more complex story; for instance, in the aforementioned subgenre of sensation fiction which partook of contemporary juridical and scientific discourses about discovery. Henry James in 1865 described Wilkie Collins's works as products of science rather than art, due to their "massive and elaborate construction [...] for the proper mastery of which [...] an index and notebook were required" (Page 1974: 123–24). In Collins's intriguing depiction of a double-purpose lecturing performance in *Armadale* (1866), authoriality painstakingly tries to assert itself, but is far from secure: dubious Doctor Downward offers a public guided tour of his Sanatorium (including its poisons and fumigating apparatuses) while secretly using his lecture to teach one member of the audience, the mischievous Lydia Gwilt, how to murder their mutual enemy. Anxiously checking on his audience's attention throughout, he betrays an awareness that his utterance necessarily exceeds his control. Adopting a patronizing tone, he brackets Lydia with Myers' "women or children [...] exist[ing] outside [science]" (1989: 172):

> "You want to see an experiment? You want to see how the little bubbles are made? Well, well! There is no harm in that. We will let Mrs. Armadale see the bubbles," continued the doctor, in the tone of a parent humouring a spoilt child.
>
> (Collins 1866: 778)

Yet this pretend lesson in science miscarries as his seemingly docile audience has secret plans of her own. Lydia's submission and attentiveness, as both audience and accomplice, are ironic stage-acts throughout (as I have argued elsewhere, the passage also implicitly comments on Collins's own metafictional manoeuvres: Zwierlein 2015: 78). Indeed, all the lecturer personae presented so far, whether historical or fictional, are engaged in variegated negotiations of power with their audiences. As Bernard Lightman argues:

> concentrating on the thoughts and methods of the popularizers does not bring us into direct contact with the audience [...]. [...] how was the message read and appropriated in different ways in different local settings by different social groups, whether they be aristocratic, middle class, or working class?
>
> (2007: 206)

Questions of authority and legitimacy in popular(izing) lectures thus cannot be reduced to Myers' simple opposition between "science" and "representing science" (1989: 172). Historians of science such as Lightman, James Secord, and Iwan Morus have underlined that scientific knowledge cannot be separated from its media of representation and performance; authority is the outcome of complex negotiations between a "professor" of knowledge and an audience (present or implied) (in addition to Lightman 2007, see Secord 2001; Morus 2011). Questions of representation and audience targeting also became more complex as the professionalization of journalism introduced standardized formats for lecture reports, from full-text versions to brief paraphrases, with specific price tags attached by the news agencies (see Meisel 2001: 270, on agencies' offers, from the mid-century onwards, of "verbatim", "full", or "summary" reports). Scientists' performance of "making visible the invisible" is, therefore, both democratic and paternalistic, and furthermore channelled through the reporting formats. "Popularization", as Hewitt has argued, is not the "dissemination" of established scientific ideas "in a diluted form" – he, too, contests the term "dilution" as used by Altick and Flanders. Rather, he proposes we view the lecture platform as an "arena of deliberation", and lecturing itself as implicated in "intellectual production" (Hewitt 2002: 1, 10, 1), with lecturers trying out ideas in the process of delivery. Nineteenth-century intellectuals like Ruskin also implicitly returned to the times of itinerant lecturers in "[taking] to the lecture as a means through which to reach an audience different from that of the periodicals" (10) and engaging in a dialogue with a community (as Ruskin did with the town of Bradford; see Hardman 1986; Morrell 1985), thus making lecturing "an important element in the endowment of knowledge production in a period before the expansion of academia" (Hewitt 2002: 22). Authority and authenticity – as well as the audience's attention – had to be produced and maintained through physical acts of embodiment. The popular, and popularizing, science lecture combined, like the demonstrations of eighteenth-century itinerant natural philosophers, instruction with entertainment, and exposition with deliberation.

Astronomical lectures: "star" performances and inattentive audiences

Victorian popular lecturing, inside or outside the British Association for the Advancement of Science or the many literary and scientific institutions, created "a platform that defined the parameters of Victorian intellectual debate that was just as effective if not more so, as the review periodicals" (Hewitt 2002: 23). In order to engage with examples of this "platform" and "debate" as narrative tropes, we will now focus on astronomy lectures in fictional and non-fictional accounts. Since the late eighteenth century, this genre had provided occasion for spectacular visual displays, with its use of lighted mechanical models of the stars and planets, three-dimensional or conveyed via large illuminated transparencies, variously termed orrery, diastrodoxon, or cosmorama (see Altick 1978: 364). Adam Walker, one of the early itinerant lecturers touring both Britain and Ireland, who "gave courses on astronomy and natural philosophy throughout the country, including the first course of scientific lectures to children at Dublin in 1781", "was renowned for his spectacular apparatus (including his giant transparent orrery)". Astronomy was frequently combined with theology, as in Frederick Grafton's 1854 "Panoramic Illustration of the Heavens", a lecture "on universal stellar creation, with thirteen transparent globes representing the planets" (Leaney 2008: 163, 178; on Walker's Eidouranion see also Altick 1978: 364). Such miscellaneous performances evoked mixed responses in the press: besides praise and admiration, there was frequently also satirical treatment. As a fixture of the Lenten season when theatres had to be closed, astronomy lectures implicitly competed with theatrical entertainments and thus acquired a reputation with some members of the public as being boring and intrusively educational (see Altick 1978: 364). In the context of a series on "Lights and Shadows of London Life", *Chambers's Journal* described on April 30, 1864 earlier times when during:

> Passion-week [...], until very lately [...] [n]othing beyond those most unexhilarating of spectacles called Orreries, and perhaps a panorama or two of the Holy Land, was offered to the metropolitan sight-seers in the way of amusement. The theatres were closed; the street-music was confined to the church-bells; and all the wicked people who wanted pleasure on Ash-Wednesday went into the country by early trains.
>
> (anon. 1864)

Indeed, views of the Holy Land, as mentioned here, were frequently displayed before or after astronomy lectures to make up slightly incongruous double bills (for some examples, see Leaney 2008: 178), and Hewitt reports that "lantern images were recycled ad nauseam, views of the Holy Land being particularly notorious for their assumed suitability for almost any topic" (2012: 84).

During the earlier stages of the century, the astronomical lecture figured in satirical treatments such as Charles Lamb's piece on the "New Schoolmaster" (1821) who was "expected to know a little of every thing" in order to offer his pupils a store of superficial information about the arts and sciences, attainments which had become fashionable among the middle classes (see Leaney 2008: 166; Howard 2004). Such "conversational" smatterings of science as markers of class distinction are the target in the *New Monthly Magazine and Humorist*, where the "superficial bore" (1838) is presented as "having qualified himself in astronomy by an attendance at two lectures on a Transparent Orrery, [...] bandies the sun, moon, and stars [like] cricket-balls; [and] regrets that he is compelled to differ from Newton concerning the principle of gravitation" (anon. 1838: 561). Scientific knowledge here has become cultural capital rather than a means to self-improvement or worshipping the Creator. Hence, the poor schoolmaster in Lamb's example "has some intrusive upper-boy fastened upon him [even during vacations ...] that he must drag after him to the play, to the Panorama, to Mr. Bartley's orrery, to the Panopticon, or into the country" (Lamb 1821; on Mr. Bartley's "Ouranologia", see Altick 1978: 364 and n.12). Here it seems the audience themselves have become "itinerants", tourists in search of shreds of information, and the satire conveys the conservative suspicion that mass entertainment has come to replace the "pure" knowledge of the chosen few. Some fictionalized lecture situations in fact dispensed with the lecturer altogether, focusing exclusively on lecture going as a middle-class fashion. An example is Sophie Prosser's novel *The Clackitts of Inglebrook Hall*, serialized in the *Leisure Hour*: in "What Happened at the Lecture", the instalment of September 10, 1870, the *nouveau-riche* but earnestly striving Clackitts have been invited by the false Sir Thomas to "a very superior lecture on astronomy [...], with diagrams and transparencies of the first order" ([Prosser] 1870: 580). His secret purpose is "to show them off – Mr. Clackitt in his ignorance, and [his daughter] Rosabella in her learning – for [...] the display of his own wit" (ibid.). Lecture going is here exclusively designed to confirm social status. Believing his guests not to have arrived yet,

Sir Thomas punctures the astronomical demonstrations with mockeries about the Clackitts' scientific "delusions" (ibid.). While the lecturer's voice is not heard once, the "Voice of Society", as Charles Dickens called his last chapter in *Our Mutual Friend* (1864–5), resounds very clearly indeed (Dickens 1864–5: 815). Mr Clackitt, who has heard everything, returns depressed, and aspiring Rosabella feels that "society had again shut its door upon her, and she was a poor, desolate outcast" ([Prosser] 1870: 580).

For its seeming concern with the other end of the social spectrum, let us now look at a double-edged, conservative perspective on the promotion of scientific knowledge to the labouring population. An anonymous contribution to the *Monthly Magazine*, published as early as 1835, pretends to report on a typical lecture "delivered in an institution dedicated to the instruction of the working community" (anon. 1835: 379). The narrator observes that the audience "were in the first place amused by watching the movements of an orrery, and then enlightened by a tissue of bombast and absurdity on the poetry of the ancients" (ibid.). He later overhears the audience response to this miscellaneous offering as he overtakes "a group of the edified class, and hear[s] the following remarks upon the lecture" (ibid.). The mechanics, in fact, are being ironical, healthily suspicious of the performance, and emphasize that more practical (if not more radical) lecture material would have been in order:

> "Why, I say, Jem, dang my buttons! But yonder fellow's a greater foo' than ony we'en had afore. My eye! Does he think we're such softies as to take his word that yon whirligig thing of his is like the universe?"
>
> "Ha! Ha! Ha!" said another, "we're no such foo's as that. I think it would ha' been more to the purpose if he'd told us how to live on six shillings a-week." In this expression I fully coincided. [...]
>
> (ibid.)

Closing with this paternalistic middle-class assent to the workingman's sentiments, this piece nonetheless expresses a common notion, traceable in Dickens's contemporaneous essays that inveigh against misplaced educational zeal directed at the working class (or children) (compare Dickens 1836: 1850). Margaret Oliphant in her later essay on "The Byways of Literature: Reading for the Million" (1858) similarly asserted, from her marked upper-middle-class position, that the "lower order[s] [...] are not to be kept in perpetual lecturedom any more than we are" (reprinted in King and Plunkett 2005: 205).

Still, the "lecturedom" of astronomy demonstrations as bestowed upon children of all social classes has been immortalized in Dickens's *The Uncommercial Traveller*, serialized in *All the Year Round*; this will serve as our last example. In the instalment of June 6, 1863, Dickens's autobiographical persona offers a satirical reminiscence of a childhood visit to "a slow torture called an Orrery" (Dickens 1863: 349). The implicit competition with the theatre, in or out of Lenten season, is established through the child's perspective: "The terrible instrument was set up at the local Theatre, and I had expressed a profane wish in the morning that it was a Play [...]" (ibid.). The apparatus is dysfunctional, the planets have holes in them, and the performance is accompanied by the monotonous lecture of a depressed exhibitioner who clearly fails to embody his message:

> All this time the gentleman with the wand was going on in the dark (tapping away at the Heavenly bodies between whiles, like a wearisome woodpecker), about a sphere revolving on its own axis eight hundred and ninety-seven thousand millions of times – or miles – in two hundred and sixty-three thousand five hundred and twenty-four millions of something elses [...].
>
> (ibid.)

While eighteenth-century orrery lectures had carried on the tradition of natural theology and catered for audiences bent upon self-improvement (see Altick 1978: 365), the formula had stopped working by the end of the nineteenth century, as an 1874 *Chambers's Journal* article recognized:

> men of science [...] fail [...] in the gilding [of the pill of instruction]. The orrery by which we are to be attracted towards the heavenly bodies, is itself a melancholy object, and only amusing from the fact that it aspires to be so.
>
> (anon. 1874: 133)

In Dickens's example, school children are the involuntary beneficiaries of a recycled astronomical lecture – a far-from-elevating experience: "when the lights were turned up again, [...] all the schools in the town [...] were discovered with exhausted countenances, screwing their knuckles into their eyes, or clutching their heads of hair" (Dickens 1863: 349). Their despair is pointedly contrasted with the pompous "official" response by an Head of School, who said that "[...] he really must beg to express his entire approval of a lecture as improving, as informing, as devoid of anything that could call a blush into the cheek of youth, as any it had ever been his lot to hear delivered" (ibid.; for Dickens's satirizing of middle-class fears about propriety, represented by the "blush [on] the cheek of the young person", see also 1864–5: 129). A recurring topic in Dickens's essays on social questions and education, the unwanted attention bestowed upon children by officious and ill-equipped educators, is here brought to a satirical edge.

Beyond the satire, the text also reminds us again of the importance of the descriptive lecture accompanying the images, which was recognized in press coverage of astronomical exhibitions, "more or less evenly apportioned between the representations of the stars and the lecturer's platform style" (Altick 1978: 364). The major part of the performance depended on the spoken word, as we have seen, because the unreliable magic lanterns "offered the prospect but not the promise of spectacle, [and] did not guarantee audience appeal" (Hewitt 2012: 84). Lantern slide sets were also used over and over again, and frequently combined with long-winded descriptions which could spell the certain doom of the entire event if delivered by a "low-spirited" performer as in Dickens's example (Dickens 1863: 349). Once again, this draws out the ambivalences of the term "popularization". As Rose confirms:

> focusing on the complexity of communication and the plurality of responses seems vital because "even where it is clear that popularizers of science sought to inculcate middle-class values and scientific optimism, it does not necessarily follow that science in popular culture functioned to that end". In fact, [...] attempts to comprehend "*lacks* of impact" could be equally significant.
>
> (2002: 247–48; quoting from Cooter and Pumfrey 1994)

In our brief survey of some journalistic lecture accounts and fictionalized versions of lectures, we have time and again recognized the importance of assessing the communicative situation as a whole, not just the "script" of the lecture. The dangers of performative "self-exposure", as well as the possibility of cross-purposes between lecturer and audience, were always present. Lively images of a "suspicious" audience, an audience not wholly subdued by a Foucauldian internalized social discipline (on recent revisions of such *grands récits* in Victorian Studies, see Zwierlein 2015: 85), but at least partly independent, "*puncturing*" the lecturer's discourse (Barthes 1971: 194), "thinking back" or about other things, were captured memorably in the nineteenth-century popular press.

From lectures to cinema: whose speech, whose images?

Contextualizing depictions of Victorian popular lectures with developments in education and rational entertainment, this chapter has singled out a few aspects for special comment: the new emphasis on a speaker's charisma and performed authority but also, since the late 1860s and the second Reform Act, a "rhetoric of public sincerity [which] came to replace the eighteenth-century notion of sympathy as the basis of [...] social coherence" (Morris 2004: 45), a rhetoric that propelled, and was conditioned by, the period's new "'star' system on the pulpit and in the lecture hall" (Hewitt 2002: 10). However, the central importance of the lecturer, "star" or otherwise, also came to be contested: not only illustrations (of stars, planets, the Holy Land) but also entire lecture scripts were increasingly recycled as a semi-professional performance business developed which fostered "enablers" behind the scenes: managers, agents, hack-writers. J. Hollingshead's "A Dramatic Collaborateur", published in *The Train* (April 1857), humorously offers the "confessions" of a ghostwriter master-minding lecture performances:

> If a new panorama is opened, I have something to do with the descriptive lecture. If Mr. Tenwigs wishes to go the round of the "Literary Institutions" with an old comic entertainment, it is my task to bring it up to the spirit of the present day; and if the same industrious gentleman should purchase a second-hand Orrery at a broker's, it is my duty to furnish him with all the latest discoveries in astronomy, that his lecture during Passion Week at one of the theatres may not be marked by any glaring errors in fact, or blemishes in style.
>
> (Hollingshead 1857: 208; see also Talbot 1876)

This hack-writer produces lecture copy to accompany science spectacles; and indeed, Altick, and after him Plunkett and Hewitt, have emphasized "the imprecision of the boundary between the popular lecture and other forms of recreational exhibition. Many of the century's dioramas or panoramas were accompanied by lecture commentaries" (Hewitt 2012: 80; see also Plunkett 2015) – not necessarily produced, we might add, by the person voicing them in front of the audience. Hewitt also reminds us of the need to economize that dominated the commercial venture of a lecture series, especially if it took place outside the major metropolitan centres: the knowledgeable lecturer embodying his own learning had become all but dispensable towards the end of the century as 'by the late 1890s, Eyre and Spottiswoode were circulating lantern lectures (written by "popular writers") with appropriate lantern slides for a small fee, promoted as a way of avoiding large lecturers' fees' (Hewitt 2012: 94). The promotion and commodification of popular performances increasingly came to be in the hands of professionals, edging out the scientist or deliberative scientific popularizer.

We could also link our inquiries into nineteenth-century mass lecture culture to Guy Debord's later questions about twentieth-century mass entertainment, the "society of the spectacle" where according to him, despite the huge crowds, communal experience had been erased. For Debord, "spectacle is [...] the development of a technology of separation. It is the inevitable consequence of capitalism's 'restructuring of society without community'" (1967/1983, par. 28). However, our examples have shown that rather than being spellbound by the performance in front of them, audiences were (depicted as) independent in thought, pursuing their own plans and intrigues, and raising their voices alongside that of the lecturer's. We have also seen that the complex negotiations between the performed authority of the lecturer and his (more rarely, her) various audiences could also result in a sheer lack of attention. As Chris Otter reminds us:

> the fact that crowds flocked to exhibitions, aquariums, and art galleries does not prove that their experience was in any way aloof, ironic, all knowing, or literary. [...] Then as well as now, a spectacle produces boredom and indifference among many, if not most, of its spectators.
>
> (2008: 7)

Moreover, in more decidedly political terms, there was a "parallel development of the culture of public lectures and the expansion of a local civic platform" (Hewitt 2002: 6). The Victorian rise of mass print culture, eyed suspiciously by political and cultural elites, thus competed against persisting cults of performance and popular orality, which were likewise variously deployed and contested in the political field. Newey also emphasizes the link with new technological possibilities: "By 1896, when the first 'moving pictures' were shown as part of music hall programmes in Leicester Square, [...] the alignment of modernity with technology in popular culture was practically complete" (2010: 156). Indeed, the lecture accompanying a visual demonstration has been seen as a precursor of the cinema (see Hewitt 2012: 80).

Let us close on a final word about the "popular" and "popularization" for the period up to 1900: James Mussell has intervened in the debate about the nineteenth-century professionalization of science by arguing that, at least where the choice of specific publication outlets was concerned, "men of science evacuated personality from their accounts of scientific discovery [while] popularizers foregrounded the personality of their narrators as a way of mediating the science for their readers" (2009: par. 3). But as we have seen, the story has to be retold with respect to the Victorian feedback loop between oral and print cultures: even "men of science" had to foreground the personal when standing in front of audiences. As soon as the voice, "both troubling and an object of desire" (Kreilkamp 2005: 32), had to be mediated in (often standardized) print accounts of oral performances, the "ownership of scientific knowledge" came to be contested (Mussell 2009: par. 3). The distinction between professionalization and popularization was fluid – and popularization was not only aiming at consumption but also implicated in the act of production. Popularizers were not marginal to scientific research; on the contrary, oral performances of science were part of the cultural process of both asserting and testing the legitimacy of systems of knowledge within the larger and conflicting discourses of the time.

Fictionalized scenes of science teaching and audience responses also implicitly comment on the paramount concerns of the nineteenth century about the new, anonymous, urban mass culture, and the processes of democratization evident in the extensions of the vote and the fledgling women's emancipation movement. Nicholas Daly's description of the nineteenth-century elitist dilemma of whether one should "attempt to capture and hold the attention of the masses", or rather to "shut out the crowd and make art, literature, and science more exclusive" (2009: blurb), can thus be reinvestigated from the perspective of mediated popular science lectures: it seems that in most cases both impulses, the democratic and the elitist, were simultaneously present. Even though lectures eventually lost their dominant place in education (see Yeo 1976: 103; Leaney 2008: 180), nineteenth-century popular lecture culture remained a complex site uniting various traditions and influences: a site where oral and print cultures were closely interdependent, where the metropolis and the provinces were joined through oral performances and press reports, where questions of legitimacy and class and gender relations were being negotiated, and where knowledge was both "popularized" and continually being produced.

Acknowledgements

This chapter was prepared in the context of a German Research Foundation (DFG) funded project on the "lecturing" of females (both active and passive) in oral and print cultures between 1860 and 1910.

References

Altick, Richard D. (1957). *The English Common Reader: A Social History of the Mass Reading Public, 1800–1900*. Chicago, IL: University of Chicago Press.

Altick, Richard D. (1978). *The Shows of London: A Panoramic History of Exhibitions, 1600–1862*. Cambridge, MA: Belknap Press.

Anon. (1835). "English Sunday Schools", *Monthly Magazine* (January): 376–79.

Anon. (1838). "A Discourse of Bores", *New Monthly Magazine and Humorist* (April): 551–61.

Anon. (1864). "Lights and Shadows of London Life: A Fast-Day". *Chambers's Journal of Popular Literature, Science and Arts*, April 30: 283.

Anon. (1874). "The Wet Blanket". In: *The Best of Husbands*, serialized in *Chambers's Journal of Popular Literature, Science and Arts*, February 28: 133–37.

Barthes, Roland (1971). "Writers, Intellectuals, Teachers [Écrivains, intellectuels, professeurs]". Trans. Stephen Heath. In: Roland Barthes: *Image, Music, Text*. New York: Hill and Wang, 1977. 190–215.

Collins, Wilkie (1866). *Armadale*, ed. Catherine Peters. Oxford: Oxford University Press, 1989.

Cooter, Roger and Stephen Pumfrey (1994). "Separate Spheres and Public Places: Reflections on the History of Science Popularization and Science in Popular Culture", *History of Science* 32: 237–67.

Crary, Jonathan (2001). *Suspensions of Perception: Attention, Spectacle, and Modern Culture*. Cambridge, MA: MIT Press.

Daly, Nicholas (2009). *Sensation and Modernity in the 1860s*. Cambridge: Cambridge University Press.

Dawson, Joseph (1896). *Peter Mackenzie: His Life and Labours*. London: C.H. Kelly.

Debord, Guy (1967/1983). *Society of the Spectacle*. Detroit, MI: Black & Red.

Dickens, Charles (1836). *Sunday under Three Heads*. London: Chapman and Hall.

Dickens, Charles (1850). "The Amusements of the People", *Household Words* 1.1 (30 March), 13–15 and 57–60.

Dickens, Charles (1863). "The Uncommercial Traveller". In: *All the Year Round* (June 6): 348–52.

Dickens, Charles (1864–5). "Chapter the Last: The Voice of Society". In: *Our Mutual Friend*, ed. Michael Cotsell. New York: Oxford University Press.

Ellegard, Alvar (1971). "The Readership of the Periodical Press in Mid-Victorian Britain: II. Directory", *Victorian Periodicals Newsletter* 13.4: 3–22.

Ellison, Robert H. (1998). *The Victorian Pulpit: Spoken and Written Sermons in Nineteenth-Century Britain.* Cranbury, NJ: Susquehanna University Press.

Featherstone, Simon (2002). "The Egyptian Hall and the Platform of Transatlantic Exchange: Charles Browne, P.T. Barnum, and Albert Smith", *Nineteenth-Century Prose* 29.1: 68–77.

Flanders, Judith (2006). *Consuming Passions: Leisure and Pleasure in Victorian Britain.* London: Harper Collins.

Fyfe, Aileen and Bernard Lightman, eds (2007). *Science in the Marketplace: Nineteenth-Century Sites and Experiences.* Chicago, IL: University of Chicago Press.

Groth, Helen (2013). *Moving Images: Nineteenth-Century Reading and Screen Practices.* Edinburgh: Edinburgh University Press.

Hardman, Malcolm (1986). *Ruskin and Bradford: An Experiment in Victorian Cultural History.* Manchester: Manchester University Press.

Hewitt, Martin (2002). "Aspects of Platform Culture in Nineteenth-Century Britain", *Nineteenth-Century Prose* 29.1: 1–32.

Hewitt, Martin (2012). "Beyond Scientific Spectacle: Image and Word in Nineteenth-Century Popular Lecturing". In: *Popular Exhibitions, Science and Showmanship, 1840–1910*, ed. Joe Kember, John Plunkett and Jill A. Sullivan. London: Pickering & Chatto. 79–95.

Hollingshead, J. (1857). "A Dramatic Collaborateur", *The Train: A First-Class Magazine* (April): 208–210.

Howard, Jill (2004). "Physics and Fashion: John Tyndall and his Audiences in Mid-Victorian Britain". *Studies in History and Philosophy of Science* 35.4: 729–58.

Huxley, Thomas Henry (1868). "On the Physical Basis of Life". In: Thomas H. Huxley: *Collected Essays*, 10 vols. New York: Greenwood. 1: 130–65.

Inkster, Ian (1981). "The Public Lecture as an Instrument of Science Education for Adults – the Case of Great Britain, c.1750–1850", *Paedagogica Historica* 80–107.

Inkster, Ian, and Jack Morrell, eds (1983). *Metropolis and Province: Science in British Culture, 1780–1850.* London: Hutchinson.

Joyce, Jeremiah (1807). *Scientific Dialogues: Intended for the Instruction and Entertainment of Young People.* London: J. Johnson.

King, Andrew (2004). *The London Journal 1845–83: Periodicals, Production and Gender.* Aldershot: Ashgate.

King, Andrew, and John Plunkett (2005). "Introduction". In: Andrew King and John Plunkett, eds, *Victorian Print Media: A Reader.* Oxford: Oxford University Press. 1–8.

Kingsley, Charles (1869). *Madam How and Lady Why, or First Lessons in Earth Lore for Children.* London: Bell and Daldy.

Kreilkamp, Ivan (2005). *Voice and the Victorian Storyteller.* Cambridge: Cambridge University Press.

Lamb, Charles (1821). "The Old and New Schoolmaster", *Elia: The Weekly Entertainer and West of England Miscellany*, July 30: 69.

Leaney, Enda (2008). "Evanescent Impressions: Public Lectures and the Popularization of Science in Ireland", *Eire-Ireland* 43.3: 157–82.

Lightman, Bernard (2007). *Victorian Popularizers of Science: Designing Nature for New Audiences.* Chicago, IL: University of Chicago Press.

McCabe, Joseph (1908). *The Life and Letters of George Jacob Holyoake.* 2 vols. London: Watts.

Meisel, Joseph (2001). *Public Speech and the Culture of Public Life in the Age of Gladstone.* New York: Columbia University Press.

Morrell, Jack (1985). "Wissenschaft in Worstedopolis: Public Science in Bradford, 1800–1850". *British Journal for the History of Science* 18: 1–23.

Morrell, Jack, and Arnold Thackray (1981). *Gentlemen of Science: Early Years of the British Association for the Advancement of Science.* Oxford: Clarendon.

Morris, Pam (2004). *Imagining Inclusive Society in Nineteenth-Century Novels: The Code of Sincerity in the Public Sphere.* Baltimore, MD: Johns Hopkins University Press.

Morus, Iwan (1992). "Different Experimental Lives: Michael Faraday and William Sturgeon", *History of Science* 30: 1–28.

Morus, Iwan, ed. (2011). "What Happened to Sensational Science?", *European Romantic Review* 22.3.

Mussell, James (2009). "Private Practices and Public Knowledge: Science, Professionalization and Gender in the Late Nineteenth Century", *Nineteenth-Century Gender Studies* 5.2; online at http://www.ncgsjournal.com/issue52/mussell.htm

Myers, Greg (1989). "Science for Women and Children: The Dialogue of Popular Science in the Nineteenth Century". In: *Nature Transfigured: Science and Literature, 1700–1900*, ed. John Christie and Sally Shuttleworth. Manchester: Manchester University Press. 171–200.

Newey, Katherine (2010). "Popular Culture", *The Cambridge Companion to English Literature, 1830–1914*, ed. Joanne Shattock. Cambridge: Cambridge University Press. 147–61.

O'Connor, Ralph (2007). *The Earth on Show: Fossils and the Poetics of Popular Science, 1820–1856.* Chicago, IL: University of Chicago Press.

Oliphant, Margaret (1858). "The Byways of Literature: Reading for the Million", *Blackwood's Edinburgh Magazine* 84 (August): 200–16. Repr. in: Andrew King and John Plunkett, eds. *Victorian Print Media: A Reader.* Oxford: Oxford University Press, 2005. 196–206.

Ong, Walter J. (1982). *Orality and Literacy.* London: Methuen.

Otter, Chris (2008). *The Victorian Eye: A Political History of Light in Britain, 1800–1910.* Chicago: University of Chicago Press.

Page, Norman, ed. (1974). *Wilkie Collins: The Critical Heritage.* London: Routledge and Kegan Paul.

Plunkett, John (2015). "Peepshows for All: Performing Words and the Travelling Showman", *Zeitschrift für Anglistik und Amerikanistik* 63.1: 7–30.

[Prosser, née Dibdin, Sophie Amelia] (1870). "What Happened at the Lecture". In: *The Clackitts of Inglebrook Hall*, serialized in *The Leisure Hour*, 976 (Sept. 10): 577–80.

Rose, Caroline (2002). "Charles Kingsley Speaking in Public: Empowered or at Risk?" *Nineteenth-Century Prose* 29.1: 133–50.

Rose, Jonathan (2001). *The Intellectual Life of the British Working Classes.* New Haven, CT: Yale University Press.

Rubery, Matthew (2009). *The Novelty of Newspapers: Victorian Fiction after the Invention of the News.* Oxford: Oxford University Press.

Ruskin, John (1865). "The Ethics of the Dust". In: *The Works of John Ruskin: The Library Edition*, ed. E.T. Cook and Alexander Wedderburn, 39 vols. London: George Allen, 1903–12. 18: 191–368.

Schafer, Simon (1983). "Natural Philosophy and Public Spectacle in the Eighteenth Century", *History of Science* 21.1: 1–43.

Secord, James (2001). *Victorian Sensation: The Extraordinary Publication, Reception, and Secret Authorship of "Vestiges of the Natural History of Creation"*. Chicago, IL: University of Chicago Press.

Sleigh, Charlotte (2005). "'This Questionable Little Book': Narrative Ambiguity in Nineteenth-Century Literature of Science". In: *Unmapped Countries: Biological Visions in Nineteenth-Century Literature and Culture*, ed. Anne-Julia Zwierlein. London: Anthem Press. 15–26.

Stafford, Barbara Maria (1994). *Artful Science: Enlightenment Entertainment and the Eclipse of Visual Education.* Cambridge, MA: MIT Press.

Talbot, Frederick (1876). "How I Came to Be a Manager", *Belgravia: A London Magazine* (February): 567–75.

Wiener, Joel H., ed. (1988). *Papers for the Millions: the New Journalism in Britain, 1850s-1914.* Westport, CT: Greenwood.

Yeo, Stephen (1976). *Religion and Voluntary Organisations in Crisis.* London: Croom Helm.

Zwierlein, Anne-Julia (2009). *Der physiologische Bildungsroman im 19. Jahrhundert.* Heidelberg: Winter Verlag.

Zwierlein, Anne-Julia (2015). "'The Subject Escapes Me': Spellbinding Lecturers and (In-)Attentive Audiences in Late-Victorian Serialized Sensation Fiction", *Zeitschrift für Anglistik und Amerikanistik* 63.1: 69–88.

12 'You ought to see my phonograph'

The visual wonder of recorded sound (1877–1900)

Elodie A. Roy

Spectacular! The phonograph as a new sensation

This chapter retraces the cultural emergence of Edison's phonographs (or 'talking machines') as they progressively became entwined with domestic and public leisure practices in the late Victorian era (1877–1900). I envision phonography and the discourses which surrounded it as sites of negotiation for modern British identities. The phonograph represented in turn a dazzling commodity, a new media, a didactic tool and an imaginary machine. Alongside other new media of the period (such as the telephone), it crystallised emerging tensions between the traditional and the new; the domestic and the global; private leisure and mass consumption. In her study of early audio in North-American households, Holly Kruse insisted that 'the phonograph was one of the earliest mass-produced, mass-marketed forms of entertainment that brought advanced technology into the home' (Kruse 1993: 3). Although it may seem an obvious point to make, one must remember that recorded sound was a *novel sensation* of the late Victorian era. As such, its status and raisons d'être were ceaselessly negotiated by both its public and its inventor, and needed to be accommodated within pre-existing and familiar routines, while effectively disrupting them. Prior to phonography, the only way of engaging with music had been through live performance; bourgeois interiors were notably populated by pianos, which women and mothers played to entertain guests or educate children. On the one hand, recorded sound questioned habitual patterns of musical consumption and progressively modified the spatial distribution of sound. On the other, it complicated the established order of listening as a natural act of perception. With the phonograph, listening became a technologically mediated experience. My claim is that phonography enhanced and 'denaturalised' hearing by drawing attention to the recordability and repeatability of sound. For the first time, one could hear one's voice, repeat the same sound twice, but also literally *see* sound and the machinery which brought it to life. It follows that phonography also irrevocably altered the perception as well as the construction of the real, particularly as it allowed for a discrete archiving of the sonic past in the form of recordings.

As noted by Asa Briggs in his minutious archaeology of Victorian things, 'the phonograph shared the headlines with the telephone (invented in 1876) as a major "topic of the day"' (Briggs 1988: 383). Accordingly, I offer a particular reflection on the role of the newspapers and illustrated papers in cementing an enduring *image* and *romance* of the phonograph, but also first as frames for the listening experience. I argue that the phonograph first reached Victorian audiences and households through the visual medium and the flourishing world of newspapers, advertisements and cartoons. Then, it may be that the phonograph was an object of visual consumption and scopic pleasure before it became the primary medium for musical consumption. Of particular importance will be the emergence of late-Victorian housewives as the primary music-buying public, and their complementary roles as curators of the modern household and mediators between 'mass-produced entertainment and the sanctity of the home' (Kruse 1993: 8).

The study of newspapers and publications of the time provided me with many valuable insights into the promotion, coverage and reception of recorded sound in British late-Victorian households. Indeed, phonographs and records were, from their first introduction to British markets, abundantly described and advertised in the press. My main sources for this chapter have been *The Times*, *The London Illustrated News* and *Punch*, three national, popular publications (*The Times* sold more than 60,000 each day and the same copy was perused by several readers) which, though different in tone, style and contents, all affected a somewhat homogeneous celebration of the phonograph. 'Certainly the most marvellous piece of mechanism the age has produced' (*The Belfast News-Letter*, 20 May 1892), the phonograph also received a sustained level of attention from local newspapers all through the period 1878–1900, as if its novelty could not quite wear out (the device was also constantly improved upon by the Edison-Bell Company during this period). My particular emphasis therefore lies with the early material culture of recorded sound; I believe that examining objects in conjunction with textual sources allows for a more dynamic and nuanced understanding of the cultural past as a multi-sensorial and heterogeneous territory. As pointed out by Kenneth L. Ames (1978) and many commentators of the period, the Victorians inhabited and busily crafted an abundant, comprehensive culture of things: few products escaped the frenzy of industrialisation and mass-production. It follows that 'it is appropriate to investigate objects that were prominent parts of Victorian everyday life precisely because the Victorians themselves were fascinated with material culture' (Ames 1978: 209). By addressing the cultural past as a multi-sensorial realm, one draws awareness to its visual, audible, tactile and olfactory qualities. Here, the sensorial can simultaneously be mapped onto the 'sensational' (Briggs 1988: 104), that which effectively offers an extension or enhancement of the ordinary senses, by means of a technology.

Bottled music: a story of Edison's cylinders

Recorded sound first entered English markets in 1878, one year after Edison patented it in the USA. The North-American public encountered the talking machine on the occasion of Edison's shows, where demonstrations would be performed on stage. The Edison Speaking Phonograph Company was established in the USA in April 1878 to '[grant] regional demonstration rights to exhibitors', by which 'individuals purchased the right to exhibit a phonograph within a protected territory' (Gitelman 2008: 31). The machine in itself constituted a spectacle and many accounts of these shows are to be found in newspapers of the time. The tin-foil phonograph was first mentioned in *The London Illustrated News* on 3 August 1877. Ten years later, the same newspaper dedicated the totality of its front page and an inside column to the phonograph, uncovering the secrets of sound-reproduction for the benefit of its curious, technophile readers:

> There is a disc of bright metal, rather larger than a shilling piece, so poised as to vibrate in correspondence with any sound that is received by the instrument. Below, and attached to this disc, is a minute point of metal, like a fine pin, which, as the diaphragm or disc vibrates, cuts an exceedingly delicate, sinuous, hair-like line into a revolving cylinder of wax. When the record is once engraved on the cylinder, we can, by reversing the movement, get back from the instrument the sounds that were put into it.
>
> (Saturday, July 14, 1888: 30)

The above passage describes a new, better phonograph which made use of wax instead of tin-foil cylinders, resulting in an improved quality of sound-reproduction. Two cartoons of Edison's invention strikingly occupy the totality of the front page, the first of them depicting 'Edison's perfected phonograph in use in the press gallery during the Handel Festival at the Crystal Palace' (a major venue for entertainment in London; see Solie 2005: 104–105). Although the festival was dedicated to the music of Handel, it is the machine and its technology-wise user which benefit from full exposure. No sign of the live performances are to be found. A further cartoon of the time, published in a French newspaper in 1897, translates well the visual appeal of the new technology: the all-mighty phonograph sits in the centre of the stage, replacing live musicians before an enraptured audience (Chew 1981:61).

Edison did not initially consider or predict the musical uses of the phonograph, but deduced a marketing list of potential uses from existing practices, which the phonograph would 'improve' upon. The first uses suggested by Edison were rather mild in scope and ambition; the idea of the phonograph as an office aid fitted well with the known fabric and economy of the Western metropolis, with its developing culture of office workers and shop assistants. In addition to this, the fact that both phonographic gear and bicycles were sold in the same shops only emphasised their status as novelty items, for which no real outlet or retailing system existed yet (because they had no stereotypical consumer as such). Music shops did primarily sell instruments and sheet music, but phonographs had to be purchased in other, recently established shops. Furthermore, the existence of phonographs in newspapers must be resituated within the contemporary trend of window-shopping; of looking, desiring and longing for commodities without ever touching them (a tendency also manifested in the extremely popular Great Exhibition of 1851 and its subsequent editions as well as smaller fairs). The gaze is an initial means of consuming the commodity and perhaps, in the case of the phonograph, the most satisfying form of engaging with it.

The phonograph, both a writing and playback device, allowed sound to be recorded on tin foil, and later wax cylinders via the means of a stylus. But it was difficult to maintain the cylinders in their original shape, despite the dedicated can in which they were stored. The early recordings were quickly dubbed 'canned music' or 'bottled music', as attested in an 1877 cartoon drawn by George du Maurier for the *Punch* almanac of the following year (*Punch's Almanack for 1878*, 14 December 1877, p. 3). The cartoon shows a man and a woman – probably servants – in the cellar of a house, replacing bottles of music (bearing labels such as Trebelli, Nilsson and Nita Gaetano, three famous opera singers of the time) on the racks. The domestic metaphor is telling: the Victorians would indeed store and accumulate possessions within the realm of the house. What had been previously conceived of as a pure, immaterial and evanescent form – music – was no exception: recording technologies allowed for music to be materialised, and consumed, amassed and circulated as any other commodity.

It can, therefore, be argued that recorded music belonged to the visual culture of the Victorians. In his influential study of Victorian things, Asa Briggs (1988) develops what he calls a 'philosophy of the eye' and argues that the nineteenth century, through cameras, microscopes and spectacles, was mainly preoccupied with the realm of seeing. Similarly, Jean-Louis Comolli described the 'machines of the visible' and 'frenzy of the visible' which marked the second half of the nineteenth century (quoted in Faulk 2004: 156). It can be argued, drawing from Walter Benjamin, that seeing was invented or reinvented through new technologies: as if the photographic processes were means of revealing the pictoriality of the word. Similarly, recorded technologies allowed people to hear for the first time; hearing their voices being played back constituted a particular and permanent shock for the Victorians. But, in order to be complete, 'a philosophy of the eye' would also need to attend to the complex emergence of the urban *flâneur*, the window-shopper, and to the new forms of visual stimulations which arose or consolidated in the nineteenth century.

Framing the phonograph: the gendered and socio-cultural location of the phonograph

A domestic and feminine machine

This section is concerned with the location of phonographic material within the domestic sphere, representing 'both the arena for and the refuge from technological penetration' (Cohen 1980: 293). It especially examines the relationship between women and new technologies in the particular instance of shopping. We should remember that the Victorian house was a gendered space, which lay under the responsibility and supervision of women. Holly Kruse (1993) reminds us that the role of the Victorian housewife was to grant or forbid access to particular commodities within the house. The phonograph, a domestic machine (Edison's first range of phonographs offered 'home phonographs'), therefore had to meet a housewife's approval before entering the domestic realm. If contemporary consumers of popular music are now more readily associated with masculinity (the figure of the male collector particularly springs to mind), the consumption of recorded music in the late Victorian age was primarily the affair of middle-class women, and the main 'purchasers of the nation's industrial products were women' (Humble 2010: 227). Reflecting upon the rise of the department store in the late nineteenth century, Rémy G. Saisselin affirms that:

> Shopping was integral to the identity of the new woman. For the first time in history the women of the bourgeoisie found themselves free and with leisure time [...]. The capital accumulation of the nineteenth century made it possible for women of this class to enjoy a certain leisure, leave their interiors, and lead a form of aristocratic life modelled on that of the old nobility.
>
> (Saisselin 1984: 39–40)

And yet, if shopping may have held an emancipatory potential, blurring the male/female spatial boundary as it 'legitimated the presence of the respectable middle-class woman on the streets of the city' (Humble 2010: 227), I would suggest that it ultimately increased the ties between women and the domestic sphere. As a matter of fact, shopping was reciprocally a way of being (physically) out of the house and (emotionally) inside it by means of the imagination: for the experience of shopping, either in the department shop or on the high street, presupposed the existence of a house to furnish. The commodities were invariably transported back to the domestic sphere and consumed – either symbolically or literally – within its premises. Early phonographs were first and foremost visual markers of musical culture and decorative artefacts in their own right before they became useful or total 'sound' devices. In this light, it may be no surprise that women should have been the ones responsible for buying what then constituted a piece of modern furnishing and a relatively superficial mark of social status, whose purchase required no particular form of expertise (see Barnett 2006).

The phonograph needed to appeal to a feminine market to gain recognition and access at home. Yet, despite their central role as curators of the house, women were not impervious to external influences or pressures emanating from husbands, male music retailers or the more invisible – yet equally male – voices of the print word which often directly spoke to them. Furthermore, it must be noted that the depiction of technology or science as female was a trope of the time; in a 1891 cartoon published in *Punch*, Miss Science, depicted under the allegorical guise of a woman, expectedly stands next to the phantom of English chemist, physicist and electric pioneer, Michael Faraday (1791–1867) listening to Edison's improved phonograph. Faraday congratulates her in the following terms: 'Well, Miss Science, I heartily congratulate you; you have made marvellous progress since my time!' (*Punch*, 27 June 1891: 309). Although men may have feminised their representation of the phonograph and of science in general, the machine was far from being exclusively reserved for female consumers in practice. For example, advertisements routinely listed uses of phonographs such as 'an invaluable assistant in letter-writing' and 'a perfect home entertainer' (*The Illustrated London News*, Feb. 25, 1899: 288). The phonograph was very often presented as appealing to younger people and those confined in the house; in an advertisement published in late December 1898, the phonograph is depicted as an ideal New Year's gift for both parents and children (*The Illustrated London News*, Dec. 31, 1898: 1012), and some models of talking-machines were miniaturised and designed for children's use (Chew 1981: 20). The talking-machine was to be used as an entertainment in the house and an administrative assistant to men in the office, where, or so the advertisement promised, 'any of your girls can write it out – shorthand not being necessary' (*The Illustrated London News*, Dec. 31, 1898: 1012). As pointed out by John M. Picker, 'the association of phonographs with women was reinforced over the 1890s by the use of the first popular commercial phonographs in the workplace, as office dictation machines, from which female amanuenses transcribed business correspondence' (Picker 2001: 772–773). It is through consumption that women coined a first narrative of modernity in the house; but this act of consumption itself, far from being autonomous, was rooted in a male dominated framework. Ultimately, women made choices with their husbands' money and arguably under the invisible (yet permanent) supervision of males.

For it to be accepted as a legitimate presence in the house or the office, the phonograph had to enhance pre-existing practices and fill in clearly known functions, pertaining to the spheres of entertainment, education or administration. The phonograph therefore mostly reinforced existing gendered roles and boundaries; it helped maintain a tradition. For example, the phonograph as a musical instrument can be linked with a longer history of musical education in the house. Bringing music into parlours had customarily been the role of women and mothers, whose role was to awaken and educate young children's tastes in music as well as teaching them to play an instrument (this was often achieved with the piano). As such, it is no surprise that the alleged educational value of the phonograph was particularly sought out (Kruse 1993: 9); in order to appeal to parents and particularly mothers, the phonograph has been advertised as being 'of material assistance to the young in their studies, and in the matter of entertaining children or guests the Phonograph is unrivalled among modern productions' (*The Illustrated London News*, Feb. 25, 1899: 288).

Yet neither the place nor function of the phonograph were sedimented as yet in the late-Victorian era; the machine equally belonged in the privacy of wealthy people's interiors and to more open, public spaces. In its early versions, the phonograph was not encased in any massive piece of furniture. Rather like an autonomous Singer pedal-driven sewing machine, it was a portative mechanical device which could be transported anywhere: 'A first-class Concert can be given in the Home, Garden, or Anywhere' (*The Illustrated London News*, April 5, 1902: 512). Indeed, as if to reflect the uncertainty as to the place of the phonograph, Edison patented several types of phonograph apart from the home phonograph (introduced in 1896); these include the Triumph (the cheapest model), the Standard and the Gem (Chew 1981: 40–41), each of them designed for a specific environment, either indoors or outdoors.

Beyond the house: popular culture and aural pleasure in the open air

Revisiting his Victorian childhood, the antiquarian F. Gordon Roe recalls a 'brass-horned affair which brayed out its music' in his uncle's country-house in Surrey (1968: 96). He also insists that 'it was rather stylish to have one of these talking machines' (97) which he also refers to as 'novelt[ies]' (96) and 'new toy[s]' (97). The phonograph was both a luxury of the better classes and a cheap popular amusement – when it was played for instance at the circus during intermission (as advertised in a programme for a day-worth entertainment at the Crystal Palace). But its status of bourgeois artefact dwindled as it progressively became an 'inexpensive' entertainment (*The Belfast News-Letter*, 20 May 1892). Prices effectively dropped steadily across the years: by 1900, the most basic of Edison's home phonographs cost £2 2s, against £6 6s only two years earlier. Other brands also started manufacturing competitive and cheaper music-playing devices at the end of the century (such as Columbia's Graphophone and Berliner's gramophones). Gramophones, introduced in 1887 and popularised in the 1890s, were almost four times cheaper than phonographs (Briggs 1988: 392). This means that more modest families were able to afford a music-playing machine, even though, more often than not, the working class encountered the phonograph outside the house, for instance in fairs or phonograph recitals, or through a proto-jukebox known as the 'coin-in-the-slot' (Briggs 1991: 46). As noted before, its existence as a popular object or as a people's object was also achieved through means of the print world (caricatures, advertisements, mail-order catalogues, sketches and photographs). In other words, the phonograph was more often than not owned and consumed collectively from a distance, as a visual landmark of the Victorian popular culture but not necessarily as a private or readily accessible good. As such, it is certain that the numbers of listeners (or potential listeners) outnumbered that of owners. It is arguably why Edison's phonograph had first been 'performed' on stage and entertained masses.

The phonograph was not necessarily or systematically being thought of as a replacement for live acts, rather, played between acts, it could offer a variety in live performances. Performances were thus hybrid, combining different types of sounds. In such a case, the phonograph was at once a playback machine and an 'instrument' in its own right, which, as soon as somebody played it, acquired a live presence of its own. The hybridity and complementarity of live performances in concert halls and domestic listening is perhaps best exemplified by the invention of the 'theatrophone' in the early 1890s. The theatrophone made use of the telephonic network to communicate or 'electrophone' live performances as they were played, allowing for subscribers (in particular customers bed-ridden by illness) to listen to live music over the phone (Briggs 1991: 45), offering a primary example of 'pleasure telephony' (85).

F. Gordon Roe (1968) remembers that the familial phonograph was mainly used to play songs by Dan Leon or Henry Lytton, two light novelty singers, as if to show that 'there was [...] no necessary alignment [...] between 'popular culture' and the working classes, and neither was there [...] an absolute divide between 'high' culture and popular' (O'Gorman 2010: 8). Indeed, popular culture is not strictly associated with a social class, but infiltrates all classes in a more uniform fashion. Victorian audiences were exposed to popular culture in a more uniform and national way 'regardless of class or gender' (Denisoff 2010:147). The combined intensification of print culture, the sustained rise of working-class wages throughout the nineteenth century and the expansion of literacy through compulsory schooling were responsible for an increasing homogeneity of culture. I would further argue that it is the shared sense of bewilderment before the machine which brought together publics from a wide range of social backgrounds, as evidenced by the unprecedented popularity of Great Exhibitions across Europe and their coverage in the press. For instance, a whole page of *The Illustrated London News* (26 October 1889, p. 536) was dedicated to 'Visitors to the Paris Exhibition Listening to the Phonograph' and offered sketches of a variety of socio-cultural publics – including children, women and men, masters and servants – listening to a phonograph side by side, together yet apart (all pressing earplugs to their heads).

Furthermore, the process of banalisation, flattening or 'uniformisation' of cultural difference was arguably enabled by the standardisation in modes of production. Only with a uniformisation of modes of production, and thereby of the standardisation of products, could an increasing homogeneity in modes of consumption and contents be achieved (Bayly 2004: 16). Early vernacular soundscapes were often characterised by novelties and ephemeral pieces, as evidenced by the programmes of music halls, which favoured 'performers who specialised in comic songs', and particularly those with a speciality in imitation and dialect (Gronow and Saunio 1999: 26). As such, idiosyncratic marks of regional identity (such as accents) travelled across the country through means of performances and recordings; but, rather than consolidating a sense of regional exception, it can be argued that recordings increased the sense of national cohesion and identity. For recordings as objects were standardised products, all alike in their external appearance; moreover, all were played with the same undifferentiated machine. It follows that the new ritual of listening to recording sound could be understood as a normative ritual, governed by its own rules and constraints.

Learning how to listen: a phonograph lesson for the kingdom

The Victorian ear as such was still untuned to recorded sound, just as the Victorian eye – one might argue – could still be deeply impressed by photographs. As Nicholas Daly points out in a piece about the Victorians and technology, 'few aspects of Victorian life escaped the impact of industrialization. The experience of time and space was transformed in fundamental ways, but so too were the visual, aural, olfactory and tactile environments' (2010: 43). What new technologies brought with them was a transformation in the lived, sensorial experience of the everyday. The experiences of phonography and photography, because of their utter novelty, were powerful and, one may argue, initially distressing; they were at once alienating, enchanting and fearful. Phonography and photography seemed to carry with them strange threats and promises of immortality, and in 1877, the phonograph was famously described in *Scientific American* in the following terms: 'Speech has become, as it were, immortal' (cited in Peters 2004: 189). Both phonography and photography (which became very popular in the 1850s) disrupted the spatio-temporal continuum, as they allowed for a past sound or sight to be conjured up again independently of the place where it first manifested itself. The phonograph arguably introduced a variety in modes of listening, and transformed listening into a potentially fractured or 'schizophonic' experience (in Pierre Schaeffer's terminology), where recorded sound is divorced from its source; that is to say, from its time and place of origin. Recording technologies allowed for a mediated or distant form of listening to occur. It represented an early shift in modes of listening and of accessing the musical world, which did not need to be encountered in the mode of the live performance anymore but could be accessed from home, at any time of the day or night. The phonograph was one amongst many other inventions that privileged the transmission of a signal across time and space, along with telephony and telegraphy. It was, moreover, what Bernard Stiegler called a 'technology of telepresence' (Stiegler 2009): it made present what had dissolved, conjuring up voices or noises which pertained to a vanished (or, in most cases, just-vanished) temporal realm. In its ideal realisation, the phonograph literally allowed one to hear the past within the present, whereas the telephone allowed one to hear sound in synchronicity, as it happened. As such, both phonography and telephony impacted upon the conception of temporality, and further shook the understanding of historical time. In reality, the phonograph was far from an ideal or perfect machine.

In order to listen to recorded sound, it may be argued that one had to literally perform a leap of faith. Listeners had to grow accustomed to recorded sound, but also learn how to listen to it; rather, they had to learn how to listen beyond the machine, for fear of only seizing its tone, noise and other idiosyncratic sounds which prevented one from hearing the recording. The poor quality of the recording needed to be compensated: seeing the machine meant hearing. Just like the early telephone, the phonograph was an imperfect media: it could not efface itself or mediate recorded sound neutrally; its mechanism could be heard. Describing the sound-marks on a tin-foil cylinder, a journalist notes that 'their tone was metallic, nasal – sometimes a squeak, indeed – very often ludicrous or miserable' (*The Illustrated London News*, July 14, 1888: 30). Furthermore, as noted by a journalist of the *Musical Times* after hearing a telephone demonstration, 'the sounds reproduced were certainly not of the sweetest kind, but the interesting and important factor was that they were there' (cited in Briggs 1988: 383). Then, the presence of sound appears to be more crucial than the quality of the signal, which seems here only a secondary or superfluous matter. In such conditions, the repertoire seems only an afterthought. Early catalogues show the heterogeneity and novelty of the recorded repertoire, and it may be suggested that the first recorded contents might have mattered less than the confounding and repeated miracle of *the machine playing*. An 1896 novelty song dedicated to the phonograph, Gus Elen's 'The Finest Flow of Langwidge Ever 'Eard', represents an exemplary case of a song stating *nothing* but its status as recorded sound (see Daly 2010: 53). The first line of the recording ('You ougt to see my phonygraff') can further be understood as a prompt to admire the machine or a narcissistic, boasting line virtually spoken by the machine itself.

The early age of recordingswas in many ways an age of innocence or naïve glorification. The bad quality of sound-reproduction was compensated by listeners' belief in sound. A phonograph, even mute, meant the presence or potential presence of music. The abundance of articles dedicated to playback devices emphasises their status as bewildering inventions. The music-player mattered more than the music it played; and the fascination with recorded sound was in fact a fascination with the technology itself. In the *Belfast News-Letter* (dated 20 Friday 1892), a phonograph recital is celebrated in the following terms: 'No more enjoyable or profitable couple of hours could be spent than with Edison's latest phonograph, which is at present on a return visit to Belfast, and can be heard every evening in the Young Men's Christian Association Hall'.

In such a context, it was not unusual for the phonograph to become a pretext for an object lesson (then one of the most widespread forms of pedagogy; see Law 2005); the showman would explain its mechanisms to the public, and the scientific dissection of the phonographic mechanical body was perhaps equally captivating as its existence. The rationality of the machine contrasts with the seemingly unaccountable miracle of the recorded voice. Phonography was more a document of sound than an art form, if art is considered, in the words of the late Victorian photographer Peter H. Emerson, as 'the selection, arrangement, and recording of certain facts, with the aim of giving aesthetic pleasure [...] so as to appeal to the emotional side of man's nature' (1899: 63). Phonography was more of an archaic, 'objective' trace of sound, even though its objectiveness was considerably flawed by the limits of the medium. The original performance had to be imagined or, at least, complemented or even reconstructed by means of the imagination. Here, imagination is literally – and liberally – understood in the broadest sense as that which conjures up visions.

Concluding remarks: the phonograph as a prophetic machine

Drawing from visual and written archival documents, this chapter has provided a retrospective reading or reconstruction of the phonographic experience which is inevitably partial. The first users have long gone, and the will to recapture a domestic, technological and emotional zeitgeist is as *naïve* as it is presumptuous.Therefore, the interpretation of material evidence remains incomplete for it cannot strictly mediate (even when the material in question is a sound-playing device) the voice of its users. It is furthermore extremely arduous to reimagine the phonograph as a new technology. This is partly because recorded sound has become 'naturalized' (Gitelman 2004: 199): it is such a prominent and banal media of contemporary Western culture that imagining a time when it was new is a difficult endeavour. Simultaneously, the contemporary proliferation of recorded sound (alongside the formidable expansion of telephony and photography) allows us to conceptualise the Victorian period as a 'laboratory of the twentieth century' (though the expression must be handled carefully). It is easy to retrospectively describe Victorian culture and its technologies as 'prophetic', introducing a clear break from previous times. What can be said for certain is that it was a crucial period of transition, which anticipated a shift from the still visual and written world to a world of leisure in motion (the dynamic and time-based form of popular entertainment which the phonograph came to epitomise can also be observed in the event of film and video-games).

What sound-recording technologies anticipated is the emergence of modern listening as a mobile and privatised listening (with headphones, a phenomenon already palpable in the theatrophone for instance). They also contributed to the more general individuation and diversification of the leisure sphere, characterised by an abundance and hyper-availability of recordings-as-commodities. But modern listening is simultaneously a private experience and an eminently homogenised one. A characteristic mark of the soundscapes of modernity and capitalism, recorded sound permeates shared spaces of work and leisure (through automated voice announcements in trains to recorded music pouring out of speakers in city centres and shopping malls). Most crucially, recorded music became one of the cornerstones of British cultural industry in the twentieth century. Recorded sound became an art *per se* (explored by sound artists, engineers, musicians), and popular genres of music directly emerged from the particular technological environment of the recording studio (see Théberge 1997). Phonography also alerted us to the power of the visual in music which was to permeate the entire pop industry. As a 1900 advertisement for the phonograph boldly put it, 'every home will sooner or later have its phonograph as a necessity'. Necessary or not, the phonograph – just like the automobile – also came to signify a desirable attribute of modern life and of the modern consumer, defined by his or her ever-increased dependency upon the latest production of capitalism.

References

Ames, Kenneth L. 1978. 'Meaning in Artifacts: Hall Furnishings in Victorian America', pp. 206–221, in Schlereth, Thomas J, (ed.) 1989. *Material Culture Studies in America*, Nashville, TN: The American Association for State and Local History.

Barnett, Kyle. 2006. 'Furniture Music: The Phonograph as Furniture, 1900–1930', *Journal of Popular Music Studies*, 18(3), pp. 301–323.

Bayly, Christopher. 2004. *The Birth of the Modern World, 1780–1914: Global Connections and Comparisons*, London: Blackwell.

Briggs, Asa. 1988. *Victorian Things*, London: Penguin Books.

Briggs, Asa. 1991. *The Collected Essays of Asa Briggs. Volume III: Serious Pursuits: Communications and Education*, New York: Harvester Wheatsheaf.

Chew, V. K. 1981. *Talking Machines*, London: Science Museum/Crown.

Cohen, Lizabeth A. 1980. 'Embellishing a Life of Labour: An Interpretation of the Material Culture of American Working-class Homes, 1885–1915', in Schlereth, Thomas J, ed. 1989. *Material Culture Studies in America*, Nashville, TN: The American Association for State and Local History, pp. 289–305.

Daly, Nicholas. 2010. 'Technology', in O'Gorman, Francis (ed.) 2010. *The Cambridge Companion to Victorian Culture*, Cambridge: Cambridge University Press, pp. 43–60.

Denisoff, Denis. 2010. 'Popular Culture', in O'Gorman, Francis (ed.) 2010. *The Cambridge Companion to Victorian Culture*, Cambridge: Cambridge University Press, pp. 135–155.

Emerson, Peter H. 1899. 'Science and Art', in Lyons, Nathan (ed.). 1966. *Photographers on Photography*, Englewood Cliffs, NJ: Prentice-Hall, pp. 60–66.

Faulk, Barry J. 2004. *Music Hall & Modernity: The Late-Victorian Discovery of Popular Culture*, Athens, OH: Ohio University Press.

Gitelman, Lisa. 2008. *Always Already New: Media, History, and the Data of Culture*, Cambridge, MA: MIT Press.

Gitelman, Lisa. 2004. 'Media, Materiality, and the Measure of the Digital; or, the Case of Sheet Music and the Problem of Piano Rolls', in: Rabinovitz, Lauren and Abraham Geil (eds). 2004. *Memory Bytes: History, Technology, and Digital Culture*, Durham, NC: Duke University Press, pp. 199–217.

Gronow, Pekka and Ilpo Saunio. 1999. *An International History of the Recording Industry*, London: Cassell.

Humble, Nicola. 2010. 'Domestic arts', in O'Gorman, Francis (ed.) 2010. *The Cambridge Companion to Victorian Culture*, Cambridge: Cambridge University Press, pp. 219–235.

Kruse, Holly. 1993. 'Early Audio Technology and Domestic Space', *Stanford Humanities Review*, 3(2), pp. 1–14.

Law, Martin. 2005. 'A Pedagogy for the Public: the place of objects, observation, mechanical production and cupboards', in: Lawn, Martin and Grosvenor, Ian (eds.). 2005. *Materialities of Schooling: Design – Technology – Objects – Routines*, Oxford: Symposium Books, pp. 145–162.

O'Gorman, Francis. 2010. 'Introduction', in O'Gorman, Francis (ed.) 2010. *The Cambridge Companion to Victorian Culture*, Cambridge: Cambridge University Press, pp. 1–13.

Peters, John Duham. 2004. 'Helmholtz, Edison, and Sound History', in: Rabinovitz, Lauren and Abraham Geil (eds). 2004. *Memory Bytes: History, Technology, and Digital Culture*, Durham, NC: Duke University Press, pp. 177–197.

Picker, John M. 2001. 'The Victorian Aura of the Recorded Voice', *New Literary History*, 32(3), pp. 769–786.

Punch. 1891. 27 June, p. 309.

Punch's Almanack for 1878. 1877. 14 December, p. 3.

Roe, F. Gordon. 1968. *Victorian Corners: The Style and Taste of an Era*, London: George Allen and Unwin Ltd.

Saisselin, Rémy G. 1984. *The Bourgeois and the Bibelot*, New Brunswick, NJ: Rutgers University Press.

Solie, Ruth. 2005. 'Music', in O'Gorman, F. (ed.) 2010. *The Cambridge Companion to Victorian Culture*, Cambridge: Cambridge University Press, pp. 101–118.

Stiegler, Bernard. 2009. *Technics and Time, 2 – Disorientation*. Redwood City, CA: Stanford University Press.

The Belfast News-Letter. 1892. 20 May, n.p.

Théberge, Paul. 1997. *Any Sound You Can Imagine*, Hanover: Wesleyan University Press.

The Illustrated London News. 1878. 3rd August 1878, p. 114.

The Illustrated London News. 1888. 14 July, p. 30.

The Illustrated London News. 1889. 26 October, p. 536.

The Illustrated London News. 1898. 31 December, p. 1012.

The Illustrated London News. 1899. 25 February, p. 288.

The Illustrated London News. 1902. 5 April, p. 512.

13 Class and the invention of tradition

The cases of Christmas, football, and folksong

John Storey

For most of human history, people have been divided into social classes, but it is only in the nineteenth century that we see for the first time the distinction between middle class and working class. EP Thompson argues that 'when every caution has been made, the outstanding fact of the period between 1790 and 1830 is the formation of "the working class"' (1980: 212). The English working class, like any class, is for Thompson 'a historical phenomenon' (ibid.). The common experience of class 'is largely determined by the productive relations into which men [sic] are born – or enter involuntarily' (9). However, the consciousness of class, the translation of experience into culture, 'is defined by men [sic] as they live their own history' (10). Therefore, it is not a 'thing', it is always a historical relationship of unity and difference: uniting one class as against another class or classes. As he explains:

> class happens when some men [sic], as a result of common experiences (inherited or shared), feel and articulate the identity of their interests as between themselves, and as against other men [sic] whose interests are different from (and usually opposed to) theirs.
>
> (8–9)

In other words, class is relational and the other group against which the working class in the nineteenth century defined itself, and against which it was defined, was the new urban middle class.

In this chapter, I will explore the entanglement of class and popular culture as it was articulated in relations between these two classes. To illustrate this, I will discuss three examples: the invention of the 'traditional' English Christmas, the development of association football and the Folk Revival. In each entanglement we find the middle class, like any other dominant class, being 'compelled ... to represent its interests as the common interest of all members of society ... to give its ideas the form of universality, and represent them as the only rational, universally valid ones' (Marx and Engels 2009: 59). This attempt to build a cultural consensus based on relations of domination and subordination is part of what Antonio Gramsci (1971, 2009) would later define as hegemony.

Inventing Christmas: Scrooge and class struggle

The 'traditional' English Christmas, 'invented' by the new urban middle class, was an invention directly connected to the processes of industrialisation and urbanisation, and was more about hegemony than it ever was about religion. Moreover, it was a creation that seemed determined to conceal the fact that it was an invented tradition. As we shall see, the new urban middle class liked to pretend they were simply continuing a tradition established by the rural gentry and aristocracy (see Storey 2010).

But how could something that is 2,000 years old be invented in the nineteenth century? The reason is simple: it did not exist, or had ceased to exist, as a popular festival in the early part of the nineteenth century. JM Golby and AW Purdue (2000) examined December editions of *The Times* between 1790 and 1836. What they discovered is quite startling in terms of the supposed 2,000 years of popular celebration of the Nativity. During this period of 47 years, Christmas is rarely mentioned and in 20 of these years, it is not mentioned at all. Further evidence of its non-existence is provided by responses to the Factory Act of 1833, which granted workers eight and a half days unpaid holiday, plus Good Friday and Christmas Day. More enlightened employers allowed workers to vote on the timing of their new holidays:

> They put it to the vote who were for Good Friday and who were for Easter Monday … . the same with regard to Christmas-Day; in our part of the country Christmas-Day is not esteemed a workman's holiday, but New Year's Day is, and the same process has been gone through of informing them that they were entitled to a holiday on Christmas-Day, and they have uniformly expressed a desire to take New Year's Day in lieu of it.
>
> (quoted in Cunningham 1980: 61–2)

The *Bolton Chronicle* reports a very similar attitude almost twenty years later in 1851:

> Not long ago the natal day of the Redeemer was pretty generally disregarded in this town, and a holiday was generally observed on New Year's Day. Now, though a holiday takes place on Christmas Day, the beginning of the New Year is looked upon as the Christmas season, and the inhabitants betake themselves to their festivities accordingly.
>
> (quoted in Hudson 1997: 115)

In some rural areas it took even longer for the new invention to take hold. As late as 1867, a book on Lancashire folklore observed: 'In some rural parts of Lancashire it [Christmas Day] is now little regarded, and many of its customs are observed a week later – on the eve and day of the New Year' (quoted in Golby and Purdue 1981: 16).

The invention of Christmas by the new urban middle class begins in the 1830s and is more or less in place by the 1880s. Everything that became part of the 'traditional' Christmas – collections of carols, cards, decorations, trees, crackers, the pantomime, the exchanging of presents, Father Christmas in the new department stores – had one thing in common: they were commodities that could be sold for a profit. Shopping more generally was foundational to the invention. Its centrality was firmly established by the 1880s, as these two examples make clear:

> A month before the event [Christmas] the fancy stores are crowded all day long with old and young in search of suitable presents, and every object is purchased.
>
> (quoted in Miall and Miall 1978: 11)

> The ... preparation for the great festival of the Christian Church ... was sounded early in November when the windows of the stationers, the bookshops, and the railway stalls became suddenly gay with the coloured plates of Christmas numbers innumerable, has increased in volume as time went on.
>
> (12)

Although the promotion of Christmas seems to get earlier each year, 'early in November' is still recognisable as at the outer limits of most people's preparations.

What the new urban middle class invented was a Christmas with a firm emphasis on commercialism and the commodity. Its central organising figure was Father Christmas and not Jesus Christ. Its guiding principle was the success of the industrial revolution. The customs that survived or were revived or invented all had one thing in common: they could be sold for a profit. Therefore, it does not make sense to bemoan the fact that Christmas is too commercial; it was invented as a commercial festival. Commodified from the very start; part of what was being celebrated was the spectacular achievements of industrial capitalism – articulated as conspicuous consumption in a market economy.

If we think of the industrial revolution as a middle-class revolution, this first part of the invention of the 'traditional' Christmas is straightforwardly a celebration of what had been achieved. However, the invention was not entirely a matter of celebration. If what was invented was commercial out of instinct, it was charitable out of a sense of fear and guilt. Charity was entangled in the invention as a direct response to class conflict, both real and imagined. The 1840s were in England known as the 'hungry forties': a period of economic slump, political struggle, and intense misery and suffering among many members of the working class (Cole and Postgate 1961). The second Chartist petition was submitted to Parliament in May 1842. Although it had 3.3 million signatures, Parliament dismissed it with contempt. Following the rejection of the petition workers went on strike in the industrial areas of the midlands, the north, and parts of Scotland, Wales, and Cornwall. The strikes were in part about resisting a reduction in wages, but they quickly became entwined with the demand for the six points of the Charter. They are usually known as the Plug Plot Riots because strikers removed the plugs from steam boilers to prevent their use. However, the term does not really capture the extent of what happened or the fact that almost 500,000 workers were on strike. Nor does it prepare us for the government's response: ten strikers killed, more than 50 deported and about 1,500 imprisoned. Mick Jenkins's (1980) study of the strikes is quite rightly called *The General Strike of 1842*. I do not think it is a coincidence that the following year is a key moment for the entanglement of Christmas and hegemony around a very particular politics of charity. December 1843 presents three forceful examples of the promotion of charity as a fundamental feature of the new invention.

The first Christmas card was produced in 1843. What it depicts is not the Nativity but a typical middle-class family eating Christmas dinner. On one side of the image of the dinner party is a group of poor people being clothed, and on the other a similar group being given food. There is a strong suggestion that the celebratory dinner is only made secure by the charity that takes place on each side of it. The same idea, made more explicitly, was expressed in the *Pictorial Times*:

> At this joyous season of dinners and laughing faces, it becomes all who are worthy to enjoy such mirth to the full, to think of the poor.... While the fire blazes on our hearth, and the table is covered so plenteously, let us think of the poor in their chilly hovels with bare tables.
>
> (23 December 1843)

In other words, as suggested by the first Christmas card, to be 'worthy to enjoy' Christmas, the middle-class reader must 'think of the poor'. Both are compelling examples of the need for charity to secure the new seasonal celebrations, but by far the most influential text at the heart of the invention of Christmas as an event organised around charity, and again appearing in December 1843, is Charles Dickens's *A Christmas Carol*. The novel did not invent Christmas, as is sometimes suggested, but it did do enormous work in popularising the invention and making material its organising ideology of charity. Like card and newspaper, the story of Ebenezer Scrooge is a warning to the new urban middle class. But to fully understand the nature of this warning, it is important that we understand what he is and what he is not. He is not a miser, as he is often represented in film and television adaptations; he is a typical member of the new urban middle class, a representative of the class as depicted on the first Christmas card and discussed in the *Pictorial Times*. When he is asked to give money for the poor at Christmas, he responds 'Are there no prisons [or] Union workhouses? .... I help support the establishments I have mentioned: they cost enough: and those who are badly off must go there' (Dickens 1985: 51). However, as the first page of the novel makes clear, he is more than just middle class; he is a middle-class businessman: 'Scrooge was ... an excellent man of business' (45). His fault is not his meanness, but his obsessive and typical focus on business. When Scrooge tells charity workers 'It's not my business' (51), he means it quite literally. Working in Manchester in the 1840s, Engels encountered many middle-class businessmen like Scrooge: 'it is self-interest, and especially money gain, which alone determines them... . All the conditions of life are measured by money, and what brings no money is nonsense, unpractical, idealistic bosh' (2009: 281–2). This is what Scrooge would dismiss as 'humbug'.

But Scrooge must learn what his former partner Jacob Marley had learned too late: 'The common welfare was my business; charity, mercy, forbearance, and benevolence, were, all, my business. The dealings of my trade were but a drop of water in the comprehensive ocean of my business' (62). He must learn to share his prosperity, again as suggested in both the first Christmas card and the article in the *Pictorial Times*. The clear implication is that if he and his class do not change, they will be destroyed. This is made very explicit in perhaps the most dramatic episode of the story, Scrooge's encounter with the two children. He is told by the Ghost of Christmas Present:

> 'This boy is Ignorance. This girl is Want. Beware them both, and all their degree, but most of all beware this boy, for on his brow I see that written which is Doom, unless the writing be erased. Deny it!' cried the Spirit, stretching out its hand towards the city.
>
> (108)[1]

The stretching of the hand towards the city is to make clear to Scrooge that these children are representatives of a much greater threat.

Ignorance and Want instead of Oppression and Exploitation is an act of political displacement on Dickens's part: education and charity rather than fundamental social change. Although charity relieves suffering, it does not change the causes of suffering. Charity is a temporary redistribution of wealth, which does not disturb the hierarchies of wealth; in fact, it safeguards them. Scrooge's 'conversion' to charity presents absolutely no challenge to the economic system that produces the need for charity. By embracing it, he will protect himself and his class from more fundamental social change. This is the conversion that he experiences: self-interest taking a more sustainable form. By learning – genuinely learning – to share a little, a great deal more will be made secured.

It is not that Dickens (and others like him) did not desire a better society, one with less squalor, less poverty, less ignorance and so on, but that a better society could never be envisaged as other than a better middle-class society. Writing five years after the publication of *A Christmas Carol*, Karl Marx and Engels reached very different conclusions about Scrooge and his class. In

The Communist Manifesto version, the boy and girl are reconfigured, not as Ignorance and Want, but as the capitalist class's 'grave-diggers':

> [T]he bourgeoisie is unfit any longer to be the ruling class in society, and to impose its conditions of existence upon society as an over-riding law. It is unfit to rule because it is incompetent to assure an existence to its slave within his slavery… . What the bourgeoisie therefore produces … are its own grave-diggers.
>
> (1998: 23–4)

As already observed, central to the new urban middle-class invention of Christmas is the claim that they are merely continuing a 'tradition'. In their discussions of Christmas, they looked back longingly to 'Merrie England', a supposed lost Golden Age in which rich and poor lived in a glorious harmony of deference and subordination. With little historical evidence, it was argued that in the feudal past 'The working man [sic] looked forward to Christmas as the portion of the year which repaid his former toils' (William Sandys, 1833; quoted in Connelly 1999: 24). Christmas was both a reward and a safety valve.

> The mirth of the humble and uneducated man received no check, from the assumption of an unseasonable gravity, or ungenerous reserve, on the part of those to whom fortune had dealt more kindly, and to whom knowledge had opened her stores. The moral effect of all of this was of the most valuable kind. Nothing so promotes a *reciprocal kindliness* of feeling as a community of enjoyment, and the bond of goodwill was thus drawn tighter between the remotest classes, whose differences of privilege, of education and of pursuit are perpetually operating to loosen it, and threatening to dissolve it altogether
>
> (quoted in Hervey 2000: 50, my emphasis).

Leaving aside the rather patronising claim that give workers a Merry Christmas and they will work happy and content for the other part of the year, we should not take these writings as serious historical accounts of the past but as fantasies intended to promote, in the present, what had never before existed. The political promise of Christmas was that enjoying Christmas together ('reciprocal kindliness') should produce a 'bond of good will', which would strengthen social relations between the classes and produce the kind of cultural consensus that Antonio Gramsci calls hegemony. Steve Roud summarises perfectly this aspect of the invention:

> In particular, [middle-class] Victorians had an uncanny knack of clothing their inventions and remodeled traditions in an aura which implied that they were traditional and ancient. They conjured up a curiously timeless Merrie England in which people were poor, but content, and

> were looked after by a benevolent local baron or lord of the manor and equally benevolent clergy. Everyone knew their place, and social hierarchies were not questioned. Such a time never existed, but in rewriting the history of our major festivals and traditions, it was this mythical Golden Age, which formed the backbone of the Victorian world-view, to which they aspired, and to which they returned again and again for ideas, ideals and legitimization.
>
> (2000: 12)

The Christmas the new urban middle class invented was both a celebration of the achievements of industrial capitalism and a temporary negation of industrial capitalism, in which goodwill replaces competition, harmony replaces conflict, and excess replaces denial. What was invented was a utopian version of industrial capitalism, in which economic competition and exploitation is softened by the temporary articulation of feudal relations of power. Moreover, it is impossible to understand the invention of Christmas without taking into account its entanglement with social class and the struggle for hegemony. The promise of Christmas is a middle-class utopia in which exploitation and oppression can exist in harmony with deference and subordination. Instead of social equality and the redistribution of wealth, it articulates the mutual obligations of rich and poor permanently bound together by the ties of charity in the best of all possible worlds.

Association football: Old Etonians or Blackburn Olympic?

My second example of the entanglement of popular culture and social class is association football. Traditional histories of football present the development of the game as passing through four stages (Dunning 1971). In its first stage, from the fourteenth to the nineteenth century, it existed as a wild and unruly game played by all social classes. The term 'football' referred to ball games that involved both kicking and handling, and may even have been used to distinguish ball games played on foot rather than on horseback. What these games had in common was a ball and the idea of getting the ball to a 'goal', but the rules were oral and various. Teams could be any size from 2 up to 2,000, and the playing area could be the whole village, or the space between two villages. A game could last all day and was often played during village celebrations (Shrove Tuesday, village fairs and feasts, etc.).

In the second stage, from about 1750 to 1840, under the pressures of the industrial revolution, the game disappeared as a popular sport. That is to say, the enclosure movement and urbanisation removed areas where the game might be played, industrialisation introduced a stricter work discipline and the new policing system enforced the law more efficiently. What remained of the popular game survived only in the universities and public schools. But even here the game was discouraged because like the game that had once existed outside these institutions, it was violent and unruly.

In the third stage, from about 1840 to 1860, the status of sport began to change: it was now seen as good for 'elite' males. Team sports, especially football, were character building, increased physical health, discipline and moral responsibility. The Clarendon Commission of 1864, established to investigate the public schools, was very clear on the benefits of sport:

> The cricket and football fields ... are not merely places of amusement; they help to form some of the most valuable social qualities and manly virtues, and they hold, like the classroom and the boarding house, a distinct and important place in public school education.
>
> (quoted in Walton and Walvin 1988: 299)

During this period, the game is supposedly civilised and codified by the public schools.

In the final stage, from about 1850 to 1890, ex-public schoolboys established the Football Association in 1863 and the FA Cup in 1871 and then, working like colonial missionaries, gradually introduced the new civilised and codified game to the working class. In an account of the development of the game, written in 1906, the author is quite clear of the role played by, in particular, Eton, Harrow, Westminster and Charterhouse: 'football, in its modern form, is entirely the product ... of the various public school games' (quoted in Taylor 2013: 22). The Wanderers were the first winners of the FA Cup. The social make-up of their team tells us a great deal about the game as played in the early days of the FA. The team included four Harrow graduates, three old Etonians, and one each from Westminister, Charterhouse, Oxford and Cambridge.

Football, it seemed, was a game intended for the middle class, but despite this it very quickly grew to become what James Walvin (2000) calls the 'people's game'. The initial challenge to the hegemony of the middle class came from Blackburn, Lancashire. In 1882, Blackburn Rovers got to the FA Cup final, losing 1–0 to Old Etonians. However, the following year, Blackburn Olympic not only reached the final, they actually won the cup, beating Old Etonians 2–1. The *Blackburn Times* (1883) understood very well how Blackburn Olympic's victory was entangled with social class:

> The meeting and vanquishing, in a most severe trial of athletic skill, of a club composed of sons of some of the families of the upper class in the Kingdom ... as the Old Etonian Club is, by a Provincial Club composed of entirely, we believe, of Lancashire Lads of the manual working-class, sons of small tradesmen, artisans, and operatives.
>
> (quoted in Walton and Walvin 1988: 299)

Blackburn Olympic's team consisted of three weavers, a dental assistant, a gilder, a plumber, a clerk, a loomer, a licensed victualler and two iron-foundry workers. A team of ex-public schoolboys would never again win the FA Cup.

From the 1870s, football as a socially organised sport developed rapidly amongst the working class of the midlands and north. Football clubs were established in different ways: through existing sports clubs (for example, Burnley, Sheffield Wednesday, Preston North End, Derby County, Notts County); promoted by religious organisations (for example, Aston Villa, Barnsley, Blackpool, Bolton Wanderers, Everton, Manchester City, Birmingham City); represented workplaces (for example, Stoke City, West Bromwich Albion, Manchester United, Coventry City, Crewe Alexandra); and by teachers and ex-pupils (for example, Blackburn Rovers, Leicester City, Sunderland).

The establishment of the Football League in 1888 was an inevitable consequence of professionalism. In order to pay wages, clubs needed reliable and regular fixtures. In 1884, Preston North End were expelled from the FA Cup because it was claimed they had used professional players. An inquiry was inconclusive, but it did discover that they had arranged jobs for players (i.e. sinecures that allowed them to be in effect full-time players). Preston got support from 40 clubs from the north and midlands. Together they threatened to form a British FA. In January 1885, professionalism was legalised. The Football League was founded three years later in 1888. Of the eleven founding teams, six were from Lancashire (Preston North End, Blackburn Rovers, Bolton Wanderers, Accrington, Everton, Burnley) and five from the midlands (Aston Villa, Wolverhampton Wanderers, Derby County, Notts County, Stoke City).

Why did the game develop so quickly in the industrial north and midlands? One compelling answer is that it had never really gone away. As we noted earlier, according to conventional accounts there was pre-industrial football, which disappeared as a popular game under the pressures of the industrial revolution. However, the public schools held on to the game, codified and civilised it, and introduced it to the world with the establishment of the FA and the FA Cup. But there is another possibility: it did not disappear but, as in the public schools, it continued to evolve in the new industrial towns and cities. In other words, the public school version was just one version, but a version with the power to impose itself on the formal organisation of the game and on the writing of the game's history. But alongside this version there existed another version we might call working-class football. The existence of this second version would also help explain how what is presented as the public school game was able to develop so rapidly in the industrial north and the midlands.

An article published in 1838 in *Bell's Life in London*, at a time when the game had supposedly disappeared as a popular sport, offers evidence for the existence of a working-class version of the game:

> A match at football will be played at the cricket ground, Leicester, on Good Friday next, between eleven (principally printers) from Derby

> and the same number of Leicester. The winners to challenge an equal number from any town in England, for a purse not exceeding £25.
>
> (quoted in Walton and Walvin 1988: 299)

In 1842, a witness at a Parliamentary inquiry into the conditions of working-class children in the mining areas of the North of England wrote:

> Although Christmas Day and Good Friday were the only fixed holidays in the mining region of Yorkshire, children had at least one day off a week and a fair portion of time in the evening. This they could use to play sport on the considerable areas of wasteland in the neighbourhood. Their games included cricket, nur and spell [a bat and ball game] and football.
>
> (quoted in Harvey 2005: 59)

Although children working long hours in the mining industry and then playing on wasteland offers little to celebrate, it does present evidence that football had continued to exist outside the universities and public schools. Therefore, athough the middle class established the FA and the FA Cup, the game's rapid development from the 1870s onwards in the industrial north and midlands suggests that the popular game had not disappeared; rather, it had simply changed in ways quite similar to what had happened to it in the public schools. Again, it is impossible to fully understand the complex history of the development of the game, and how this history has been written, without including as part of the explanation the important role played by social class.

The Folk Revival: fakesong and the imagined village

The Folk Revival of the late nineteenth and early twentieth centuries offers perhaps the most compelling example of the entanglement of class and popular culture. It was a revival in which the role of the actual folk – the rural working class – was mainly symbolic. They were the mere carriers of something they did not really understand. Fortunately, the middle-class collector was at hand. The honest and cheery peasant – the rural working class of the middle-class imaginary – joyously and deferentially welcomed their intervention:

> Imagine, then, the joy when the collector calls upon them and tells them of his love of the old ditties. He has only to convince them of his sincerity to have them at his mercy. They will sing to him in their old quavering voices until they can sing no more; and, then, when he is gone, they will ransack their memories that they may give him of their best, should, perchance, he call again, as promised.
>
> (Sharp, 1907: 106)

Reverend Sabine Baring-Gould's account of his last encounter with 'Old Hard', a disabled stone-breaker, living with his wife on £40 a year, reveals a great deal about the class relations of folksong collecting: 'Then and there I obtained from him a further crop of ballads. That was the last reaping, for in the ensuing bitter frost the aged man was found dead, frozen on a heap of stones by the roadside' (quoted in Harker 1985: 159). In this drama of nature and culture, Old Hard is not even human: he is presented as an aspect of nature, a field from which Baring-Gould, the man of culture, will never again be able to reap a harvest. Percy Grainger's methods of collecting are symptomatic of the general process of appropriation. As Georgina Boyes points out:

> Percy Grainger's tricking an elderly woman into singing songs he had earlier been refused, by hiding under her bed and recording the results of a request for them from her grand-daughter, cannot easily be characterized as anything but ruthless expropriation.
>
> (2010: 49)

Taking Cecil Sharp as the classic example of the middle class collector, Dave Harker observes that:

> No matter what working men and women sang, loved, or treasured, now that the products of their culture had been traded for a mug of cider, a quid of tobacco, a few pounds or some other trinket, they were Sharp's property, to do with as he thought fit. They were [now] a commodity with a market value, an ideological 'instrument' and cultural 'raw material' for the class of which Sharp was … [a] member.
>
> (1985: 197)

The rewards of appropriation were often considerable. As Boyes points out:

> In 1910, Sharp's income from full-time promotion of the Revival was £536 – £373 of which was from royalties and £161 from lecturing. By 1922, he was earning enough from his publications and lecturing to consider it appropriate to relinquish his Civil List Pension.
>
> (2010: 50)

In other words, Sharp was earning almost 11 times more than the people he typically defined as the folk – a rural labourer would often earn less than £50 a year. Similarly, songs taken from the elderly in workhouses earned royalties that would have been a welcome income to such inmates, had they been offered this as an alternative to token payments in money or in kind.

Once collected, what had been the communal property of the rural working class was now the private property of middle-class collectors. This was confirmed in the Copyright Act of 1911. This was appropriation rather than preservation. It was a little like a cultural equivalent to the enclosure movement. What had once been in common ownership was now private property. The songs of working people were taken, edited and used, to the profit of the collector, to produce a renewed middle-class culture. The ways in which the material collected was re-worked should make us very sceptical of all claims of preserving an authentic tradition. As Boyes points out, 'how far what was presented as "folksong and dance" represents the actual repertoires and performances of the turn of the century rural working class ... [is] highly problematic' (47).

By ideological sleight of hand, the middle-class collector became the folk. Taken from the rural working class, once censored and rewritten, folksong became a part of middle-class culture. In order to avoid what he called the 'obscene or indelicate', Baring-Gould wrote 'fresh verses' (quoted in Harker 1985: 166). His aim was to produce songs that were 'tolerable to men and women of culture' (ibid.). He felt able to do this because he believed he was 'imbued with the feeling of the folk poet' (ibid.). The tastes and assumptions of the middle-class collectors were treated not as socially conditioned and historically situated, but as universal. What had supposedly evolved for hundreds of years was now fixed as if history had come to an end. In an article in *Folk-Lore* (1893), Joseph Jacobs was quite open about the suspect conceptual solidity of 'the folk'. As he pointed out: 'The folk is simply a name for our ignorance' (quoted in Harker 1985:170). Ignorance is one way to understand it, but I think we can also see it as a form of a fantastical class substitution. The folk are used to disguise what is a middle-class appropriation and invention, producing what Harker (1985) calls 'fakesong', emanating from what Boyes (2010) calls 'the imagined village'.

In his inaugural address to the Folk Song Society in 1899, Sir Hubert Parry makes very clear that more than music was at stake. He sees in folksong the possibilities of a new consensus in which divisions of class would lose their antagonistic qualities and exist only as differences between people who shared a fundamental Englishness. But he warned his audience that 'there is an enemy at the door of folk music which is driving it out, namely the common popular [music hall] songs of the day; and this enemy is one of the most repulsive and most insidious' (1899: 1). Music hall is thus presented as symptomatic of the supposedly degraded culture of the urban working class. As Parry explains:

> If one thinks of the ... terribly overgrown towns ... where one sees all around the tawdriness of sham jewellery and shoddy clothes ... [and] people who, for the most part, have the most false ideals, or none at all ... who think that the commonest rowdyism is the highest expression of human emotion; it is for them that the modern popular music is made, and it is made with commercial intention out of snippets of musical slang. (1–2)

The problem, however, was not just that music hall was so bad, but that it:

> will drive out folk-music if we do not save it. For even in country districts where folk-songs linger, the people think themselves behindhand if they do not know the songs of the seething towns; and as soon as the little urchins of distant villages catch the sound of a music hall tune, away goes the hope of their troubling their heads with the old fashioned folk-songs.
> (2)

However, the reintroduction of folk music to the working class has potential benefits that reach well beyond the singing of songs. Parry told the Folk Song Society:

> [we can] comfort ourselves by the hope that at bottom, our puzzling friend, Democracy, has permanent qualities hidden away somewhere, which may yet bring it out of the slough which the scramble after false ideals, the strife between the heads that organise and the workmen who execute, and the sordid vulgarity of our great city-population, seem in our pessimistic moments to indicate as its inevitable destiny.
> (3)

In other words, folksong was a cause for political optimism. Cecil Sharp makes a similar argument:

> Flood the streets ... with folk-tunes, and those, who now vulgarise themselves and others by singing coarse music-hall songs, will soon drop them in favour of the equally attractive but far better tunes of the folk. This will make the streets a pleasanter place for those who have sensitive ears, and will do incalculable good in civilising the masses.
> (1907: 137)

In particular, Sharp advocated the introduction of folksongs into elementary schools as a means to 'civilise the masses'. He was in no doubt of their 'value as an educational force' (134):

> For, good music purifies, just as bad music vulgarises; indeed, the effect of good music upon minds of children is so subtle and so far-reaching that it is impossible to exaggerate the harmful influence upon character which the singing of coarse and vulgar tunes may have. Up until now, the street song has had an open field; the music taught in the schools has been hopelessly beaten in the fight for supremacy. But the mind that has fed upon pure melody of the folk will instinctively detect the poverty-stricken tunes of the music-hall, and refuse to be captivated and deluded by their superficial attractiveness.
> (135)

Folk culture was a potential source of national unity. It offered the possibilities of a shared culture in which the 'workmen who execute' would once again be 'cheery' and 'content'. An account of a pageant held in Ripon, Yorkshire in 1886 is indicative of attempts to use folk culture to build a very particular form of consensus: 'The distinction of class of course could not be disturbed, but class prejudice could be broken down, and those gatherings that had been held during the past week were eminently calculated to do that' (quoted in Boyes 2010: 35). This is classic hegemony: class continues to exist as difference drained of antagonism.

In many ways, the idea of folksong was a middle class fantasy, constructed through denial and distortion. It was a fantasy intended to challenge the perceived dangers of the present, and safeguard the future. It represented a lost world of the authentic, a culture before the fall into industrialisation, urbanisation and the awful visibility of antagonistic class relations. The middle-class collectors of folksongs idealised the past in order to condemn the present. The rural worker – the peasant – was mythologised as a figure of nature, a 'noble savage' cheerfully walking the country lanes and working without complaint in the fields of his or her betters – the living evidence of, and a link back to, a purer and more stable past.

The pastoral make-believe of the folk offered an alternative to the rather troublesome spectre of the urban-industrial working class. In this sense, then, the intellectual cult of the folk-singing folk was a nostalgic fantasy of a time when working people recognised their inferiority and acknowledged due deference to their social superiors. As John Carey observes of intellectuals of this period, they 'preferred peasants to almost any other variety of human being, since they were ecologically sound, and their traditional qualities of dour endurance, respect for their betters and illiteracy meant that the intellectual's superiority was in little danger from them' (1992: 105). Parry praised 'the quiet reticence of our country folk, courageous and content, ready to meet what chance shall bring with a cheery heart' (1899: 3). This is what the working class had once been – 'what lovely qualities there are in unsophisticated humanity' (ibid.) – and it offers reassurance to its middle-class collectors that what had been might be again.

The songs of the folk allowed middle-class intellectuals to imagine a lost national and natural identity, and to dream of the possibility of a new 'authentic' national unity of a people bound together once again by the organic 'ties of land and language' (Martin-Barbero 1993: 12). At the centre of this fantasy stood an image of 'the people' that resembled neither the rural or urban working class. What resulted, as this empty, impossible category was valorised, was an active denial of the actual lived cultures of working people, both rural and urban – what the Colombian theorist Jesus Martin-Barbero has described as 'abstract inclusion and concrete exclusion' (7).

Conclusion

What these three examples should have made very clear is that popular culture and social class were in the nineteenth century deeply entangled, and that our understanding of these and other examples of popular culture will remain incomplete until we critically engage with the ways in which social relations of class structure their emergence and conditions of possibility. In each case, we see the playing-out of hegemony, as the middle class seek to establish a consensus which secures for itself a leading position in relation to the working class. In other words, to return to the quotation from Marx and Engels, we have in each example the playing out of the middle class's need to 'represent its interests as the common interest of all' (Marx and Engels 2009: 59).

I am conscious of the fact that I have stressed incorporation and ignored resistance, which certainly existed and still exists, but England remains a country in which the homeless are taken care of at Christmas and then returned to the streets, football may well be the 'people's game' (Walvin 2000) but it is still run by the middle class, and folk music continues to be (for many on both the left and right) the supposed standard against which to measure negatively all other forms of popular music.

Note

1 John Leech's illustration 'Ignorance and Want' from *A Christmas Carol* (1843) can be viewed here: www.bl.uk/romantics-and-victorians/articles/the-origins-of-a-christmas-carol

References

Boyes, Georgina, 2010. *The Imagined Village*, Leeds: No Masters Co-operative Limited.

Carey, John, 1992. *The Intellectuals and the Masses*, London: Faber and Faber.

Cole, GDH and Postgate, R, 1961. *The Common People*, London: Methuen.

Connelly, Michael, 1999. *Christmas: A Social History*, London: Tauris.

Cunningham, Hugh, 1980. *Leisure in the Industrial Revolution*, London: Croom Helm.

Dickens, Charles, 1985. *A Christmas Carol*, Harmondsworth: Penguin.

Dunning, Eric, 1971. 'The Development of Football'. In E Dunning (ed.), *The Sociology of Sport*, London: Frank Cass, pp. 133–151.

Engels, Frederick, 2009. *The Condition of the Working Class in England*, Oxford: Oxford University Press.

Golby, JM and Purdue, AW, 1981. 'Christmas'. In *Popular Culture* U203, Milton Keynes: Open University Press.

Golby, JM and Purdue, AW, 2000. *The Making of the Modern Christmas*, Stroud: Sutton.

Gramsci, Antonio, 1971. *Selections from Prison Notebooks*, London: Lawrence & Wishart.

Gramsci, Antonio, 2009. 'Hegemony, Intellectuals and the State'. In J Storey (ed.), *Cultural Theory and Popular Culture: A Reader*, London: Routledge, pp. 85–91.

Harker, Dave, 1985. *Fakesong: The Manufacture of British 'Folksong' 1700 to the Present Day*, Milton Keynes: Open University Press.

Harvey, Adrian, 2005. *Football: The First Hundred Years*, London: Routledge.

Hervey, Thomas K, 2000. *The Book of Christmas*, Hertfordshire: Wordsworth.

Hudson, John (ed.), 1997. *Dickens' Christmas*, Stroud: Sutton.

Jenkins, Mick, 1980. *The General Strike of 1842*, London: Lawrence & Wishart.

Martin-Barbero, Jesus. 1993. *Communication, Culture and Hegemony*, London: Sage.

Marx, Karl and Engels, Friedrich, 1998. *The Communist Manifesto*, Peking: Foreign Languages Press.

Marx, Karl and Engels, Friedrich, 2009, 'Ruling Class and Ruling Idea'. In J Storey (ed.), *Cultural Theory and Popular Culture: A Reader*, London: Routledge, pp. 68–69.

Miall, A, and Miall, P, 1978. *The Victorian Christmas Book*, London: Dent.

Parry, Hubert, 1899. 'Inaugural Address'. *Journal of the Folk Song Society* 1:1, 1–3.

Roud, Steve, 2000. 'Introduction'. In TK Hervey (ed.), *The Book of Christmas*, Hertfordshire: Wordsworth, pp. 9–13.

Sharp, Cecil, 1907. *English Folk-Song: Some Conclusions*, London: Novello.

Storey, John, 2010. *Culture and Power in Cultural Studies*, Edinburgh: Edinburgh University Press.

Taylor, Matthew, 2013. *The Association Game: A History of British Football*, London: Routledge.

Thompson, EP, 1980. *The Making of the English Working Class*, Harmondsworth: Penguin.

Walton, John K and Walvin, James, (eds), 1988. *Leisure in Britain 1780–1939*, Manchester: Manchester University Press.

Walvin, James, 2000. *The People's Game: The History of Football Revisited*, Edinburgh: Mainstream Publishing Company.

14 Ripping yards

Capturing (not catching) and constructing the myth of Jack the Ripper in nineteenth-century London

John Paul Green

Judith Flanders' summary of the events and names attached to the Whitechapel murders of 1888 perfectly illustrate the ways in which Jack the Ripper is immediately rendered as both fictional and 'real':

> While Sweeney Todd never existed, and the 1888 Whitechapel serial killer definitely did, everything we know about Jack the Ripper – his name, his persona, his reasons for killing – is the culmination of a century of murderous entertainment, of melodrama, of puppet shows, of penny-dreadfuls and more.
>
> (Flanders, 2011: 465)

The moment we begin to discuss the Whitechapel murders we are thrown into a fog of both fact and fiction. 'Real' and 'facts' immediately become interchangeable and problematic terms, given the overall lack of evidence, contradictory reports and accounts and, ultimately, lack of perpetrator in the case of the Ripper murders. Little can be gained in fact by treating the crime and criminal as an historical and factual moment, and should instead be looked upon as part of a wider series of issues relating to the making of modern popular culture, an industry as artificial and sensational as stage melodrama or early crime fiction.

If the wealth of books available and continuing to be published on the 'definitive' or 'complete' accounts of Jack the Ripper in recent years can be used as a gauge, there is little sign that interest in Jack the Ripper and the associated Whitechapel murders has in anyway diminished since the first newspaper accounts in 1888. If anything, the mini-industry that arose immediately after the initial murder, commonly attributed to the Ripper, has itself contributed to a much broader commercial venture, that of real-crime reportage in the form of biographies of twentieth-century murderers – the likes of John Christie, Ed Gein, the Moors Murderers, Peter Sutcliffe ('The Yorkshire Ripper'), Fred West and Harold Shipman have all followed in the footsteps of the Ripper in terms of the marketing and commodification of notorious murderers and serial killers. In discussing the continued appeal of the Ripper and his crimes, Gary Coville and Patrick Lucanio assert that the Ripper 'has transcended mere criminality to become a representation of the dark side of ourselves, and hence the Ripper is no longer somebody, but everybody' (1999:8). I would move beyond this introspective psychoanalytical reading, however, and suggest that the Ripper represents our fascination not only with ourselves but our sense of time and place within the context of crime. Furthermore, Jack the Ripper and the Whitechapel murders can be read not only as part of the same criminal event, fixed in one temporal space (as an historical moment in 1888) but that the Ripper also exists successfully beyond this period (in fact, both temporally and spatially in some cases), through various fictions including several examples from the science fiction genre which provide interesting adaptations on the Ripper, including but not limited to *Star Trek's* 'Wolf in the Fold' (1967), *Babylon 5*'s 'Comes the Inquisitor' (1995) and the film *Time After Time* (1979), the latter example managing to bring together the Ripper, H.G. Wells (and his time machine!) and Sherlock Holmes in a modern-day San Francisco. Such examples are what Clive Bloom et al. refer to as 'the imagination of the future' (1988:136), where fictional and factual characters exist beyond their original texts or historical moments. In addition, the Ripper and the crimes continue to exist in the so-called real crime books that claim to be the definitive work on the subject. The Ripper plays on anxieties of crime; more specifically, violent sexual crime at any given time. These anxieties, as this chapter will discuss, were both exploited and consumed, by a variety of sources, from the British press through to the literature and popular entertainments of the time. Coville and Lucanio have observed that 'literary parodies and pastiches of the Ripper... dramatically escalated over the intervening years at a rate in keeping with the expanding power and diversity of popular entertainment' (1999:9), but while this is undisputedly true, the Ripper and the crimes did not create this industry, but were themselves exploited, very successfully, by a pre-existent industry in 1888.

Since the late nineteenth century, the reporting of violent sexual crime has been elevated not only to the headlines in the national press, but now fills the shelves of high street book stores and supermarkets, providing interested parties with a wall of real crime from which to choose. These real accounts sit alongside fictional detective novels and thrillers, and it is in-between the two sub-genres that Jack the Ripper sits, blurring the lines between fact and fiction. He inhabits the 'real' Whitechapel, where the actual murders took place, but he also exists in a mythical Whitechapel, the fog-shrouded, poorly gas-lit streets, popularized in nineteenth-century literature, gothic stage melodrama and later in early British and Hollywood cinema. Guy Barefoot, in his study of cinematic gaslight melodrama, draws on the work of Colin McArthur to observe that the early cinematic renderings of fog-bound London themselves drew on 'traveller's accounts of London, the novels of Charles Dickens... the Sherlock Holmes stories of Sir Arthur Conan Doyle, and press accounts of the Jack-the-Ripper murders' (2001:57). The

Ripper exists as comfortably (if not more so) with fictions as he does within historical realism.

Hundreds of books have been published claiming to reveal the true identity(ies) of the Ripper, based on a wealth of police documents, newspaper reports, hearsay and rumour. Each year a new theory is proposed, supported by 'new' evidence or reinterpretation of existing documents. Ultimately, however, we are no closer to finding the 'truth' simply because the 'evidence' is predicated on a series of untruths, fictions and unsubstantiated material. The closer the writers seemingly come to the truth, the further the myth is established. To appropriate Jean Baudrillard's assertion:

> We live in a world where there is more and more information, and less and less meaning... a sphere of information that is purely functional, a technical medium that does not imply any finality of meaning... [but] something else that in a sense comes after the fact.
>
> (1994: 79)

That is to say, the more we 'know' about the Ripper, the further from 'truth' we get. Furthermore, if we apply Baudrillard's second hypothesis from 'The Implosion of Meaning in the Media' that 'the extent of information is directly destructive of meaning and signification, or that it neutralizes' (ibid) meaning, we can also argue that the abundance of information further suggests that the more so-called 'definitive evidence' is provided, the more Jack the Ripper is placed in the realm of fiction and myth. Here we can ascribe Roland Barthes' work on myth, where the continued conspiracies surrounding the Ripper murders act as 'a conjuring trick' that have 'turned reality inside out... emptied it of history' (in Storey, 2009:269). Therefore, to *know* or want to know the Ripper as an historical person of fact is ultimately a futile exercise. The Ripper and the crimes exist most successfully as a series of fictions or narratives that have been written and adapted, translated and reappropriated since 1888. As Alexandra Warwick notes, the actual murders 'fit the motion of serial publication: sequential events recognizably part of the same work, varying enough to maintain interest' (2007:72). Furthermore, the use of the slogan 'Autumn of Terror' brands the murders, placing them in a mythical time as well as space. In addition, we have the description of the murders of Elizabeth Stride and Catherine Eddowes, dubbed the double-event by the press, a macabre rebranding of the murders, turning a tragedy into sensation and spectacle.

In common with the work of Alexandra Warwick and Martin Willis, I do not seek to make a fiction or a frivolity of the fact that several women were brutally murdered during a relatively short period of time. It is not the 'denial of the reality of the grotesque deaths of five or more women' (2007:2), but to appreciate the interest and popularity of the crimes in 1888 as well as into the twentieth and twenty-first centuries, we need to move beyond the actual murders as a crime and examine the whys and ways in which these killings became a sensational event, one which captured the public's imagination and generated a mini-industry of its own. Consequently, this chapter is also not concerned with unmasking of the killer. To do so would be both futile and irrelevant. The allure of the crimes and the Ripper as a figure lie in the fact that they exist in myth – in a series of oft-told and adapted stories, from stage to screen, from news print and songs. The intention here is to offer ways of engaging with and 'reading' the Ripper 'stories' in the context of their immediate popularity in the late nineteenth century and to make sense of their continuing popularity today. Jack the Ripper and the Whitechapel murders quickly became a cultural commodity to be packaged and sold by media industries that had previously drawn inspiration and commercial success from stage melodrama, penny dreadfuls, horror literature as well as the continuing success of regional and national press and lifestyle magazines such as *Titbits* (1881–1989) and later *The Strand* (1891–1950), which continued to trade on the unsolved mystery and almost inexhaustible list of suspects in the years during and after the 1888 crimes.

The chapter will provide three ways of reading the Ripper in terms of his popularity in the late nineteenth century. First, I will examine the role of the British press in creating, propagating and cementing the myth of the Whitechapel murders primarily as a way of selling newspapers. Here I will examine the ways in which the press was drawing on existing forms to report crime and to provide and satiate audience appetite for the grotesque and macabre. Second, the chapter will examine the broader cultural context of the Ripper in terms of a continued tradition of an interest in gothic horror and sensation. Here I will examine the role of other popular fictions in feeding the myth of the Ripper, or indeed responding to the crimes in their own creation. I will position the Ripper within a heritage of penny dreadful fictions (chiefly Spring-Heeled Jack and Varney the Vampire), the birth of the horror and crime novel and the popularity of stage melodrama, notably gothic melodrama. The third approach will be to examine the ways in which the Whitechapel murders and the figure of Jack the Ripper were used as shorthand to discuss anxieties around class and poverty within East End London, where the Ripper can be read as both symptomatic of the environment and potentially its saviour in the revealing to the general public of such poor social conditions in inner-city London. Finally, the chapter will look beyond the nineteenth century to highlight the influence the Ripper industry, created in the 1880s, has had on contemporary popular culture.

The press

> The newspapers had been the most notable element of the entertainment industry's response to the crimes, because of the quantity of their coverage.
>
> (Flanders, 2011:452)

Judith Flanders' *The Invention of Murder* is clear in its assertion that modern crime was an invention of the Victorian age, seeing the rapid escalation of crime reporting and the selling of sensation in magazines, books and theatre and part of an increased commercialization of popular culture. While her work doesn't focus on Jack the Ripper, it is clear, that the Whitechapel murders enabled the London (and later national) press to fully exploit an interest in death, murder and detection. She states that 'the series of deaths which came to be grouped together as murders by a single, unknown person, were, without doubt, a circulation-booster' (2011:425). While early news accounts focused on the barbarity of the crimes (themselves not unique), it was the tying together of a pattern and a killer that helped fuel interest in the killings. As soon as a name was given to the murders, the reporting and popularity of the crimes intensified. If the Ripper was 'given birth to' he was surely 'born' in a press office. Certainly, the infamous moniker 'Jack the Ripper' was first seen on 27 September 1888 in a letter received at the Central News Agency. This correspondence became known as the 'Dear Boss' letter, signed for the first time by a 'Jack the Ripper'. Various Ripperologists have suggested that this was a hoax, created by either a newspaper editor or journalist. While the provenance of the letter has yet to be confirmed, the fact is that from this moment, the name Jack the Ripper became the brand name for the Whitechapel murders, a brand that was utilized significantly by the London press.

Darren Oldridge's work on the role of the British press in capitalizing on and exploiting the Whitechapel murders is useful in exploring the ways in which newspapers tied together the murders to form an event. Oldridge identifies that little connected the victims, save their profession and 'tragic lives', and suggests, rather persuasively, that the London regional press, chiefly *The Star* and *Pall Mall Gazette*, were instrumental in propagating myths dressed as stories to a public hungry for sensation and horror: 'By crafting the story of the Whitechapel murder… the press launched the 'Ripper Industry' on an enthralled public' (2011:54). Here we see the regional press not only feeding an audience hungry for dark drama, but also fuelling a moral panic about the state of the East End, namely the Whitechapel area.

L. Perry Curtis' definitive work on the Ripper and the British press offers an illuminating examination of the role the London newspaper industry played in developing and marketing the image of Jack the Ripper, as well as the ways in which the murders were reported in more general terms. Curtis provides an insightful overview concerning the popularity of not only the Whitechapel murders, but the relationship between the newspaper industry and its readership:

> If murder is a social (as well as antisocial) act, then its telling and selling by the press are significant cultural events that reveal much about what journalists think the public wants or needs to know. Murder news by

> definition both whets and feeds an appetite that disapproving critics deem perverse or voyeuristic.
>
> (2001:9)

On this basis, the conditions for the reporting of the Ripper killings were ripe for exploitation, to both an industry selling crime and a readership purportedly hungry for it. Under such conditions, it is understandable that the solving of the crime meant less to the newspaper owners and editors than its continuation.

More broadly, Rosalind Crone discusses the development of the popular press in the 1800s, where 'during the first half of the nineteenth century, the broadside trade dramatically expanded to become a dominant method of crime reporting for the common people and a central feature of life in London' (2012:96). She notes also 'no details of violent crime were considered too shocking or distasteful for publication' (2012:101), and that being respectable 'mattered very little to profit-hungry news proprietors when collecting and publishing reports on gruesome murders' (2012:249). In this respect, *The Star* and *Pall Mall Gazette* were simply continuing a tradition of detailing, quite explicitly, the violent crime committed in inner-city London. It is these moral panics around London, notably the East End, that were countered by Arthur Conan Doyle's detective, Sherlock Holmes, who provided a way of not only 'reading' the labyrinthine and ill-lit East End, but also dispelling the myths that violent crime occurred on such a grand scale as the local press would have its audience believe. It is not that the Whitechapel murders in themselves created this type of news reporting; rather, the crimes provided a rich vein of material from which the press could tap.

These accounts fed into a general concern, if not also fear, of 'outcast London', whose very geography and slum housing, according to Gareth Steadman-Jones, 'was suffused with all the trappings of a romantic horror story... intensified by a spate of criminal novels... [but] beneath the romantic trappings lay the fear of a hardened semi-criminal race of outlaws' (1984:179–180). Certainly the location of the so-called Ripper murders helped fuel the myths and fears of the East End and as Michael Diamond contests, pushed the stories towards the sensational (2004:184), partly due to racial tensions being stirred up as a consequence of the significant number of Polish and Russian Jewish immigrants in the area, a situation not unlike we have seen in recent years with the right-wing press and fears of European immigration and Islamophobia. The location, along with the name of the suspect, became part of an entertainment industry keen to exploit audience fascination with the crimes, and walking tours (still popular today) were conducted around the sites where the Ripper's victims were discovered during 1888. Newspaper illustrations of the time show how crowds would gather at the murder sights, some to protest at the crimes themselves (or the police force's inability to catch the killer), while others were seen as spectators to the crimes, which themselves were sold as events as part of the 'Autumn of Terror'. On November 17 1888, *The Illustrated Police News* further added an element of horror in its coverage of the murder of Mary Jane Kelly, referring to the murderer as 'the monster of the East End'. The Whitechapel area of London was being written as a gothic horror story, and Jack, whose name, as Coville and Lucanio suggest, implies the everyman as well as trading of previous criminals such as Spring-Heeled Jack and Slippery Jack (1999:8), became the lead villain.

London as stage

If the London press was playing out the gothic horror story in the reporting, then the East End of London can be seen as its stage. If Conan Doyle's Sherlock Holmes (published since 1890) can be seen to offer a way of opening up the city, making the unseen visible and offering a clear (albeit fictional) pathway through the streets of London, alleviating some of the middle-class anxieties that parts of London had become a haven for the dangerous classes, infecting those around them, it did so to provide an antidote to the fears being played out earlier during the Ripper killings as well as, more broadly, concerns about the living conditions of the poor in the East End. John Springhall's work on the nineteenth-century penny dreadfuls suggests that many of the stories failed to engage with the harsh realities of inner-city London life, and instead continued the traditions of gothic melodrama in placing the villainy squarely at the foot of aristocracy in a romanticized and often rural setting. For Springhall, the penny dreadfuls avoided '...the real conditions of East London's huge casualized labour force, characterized by low wages, irregular employment and foreign immigration, are never directly confronted in penny fiction' (1998:67). He continues to argue:

> the popular literature of seduction evaded the reality of sexual assault within the working class in Victorian times... This kind of writing also contributed to a wider rhetoric of the poor as weak and passive victims who needed to be protected by kindly middle-class paternalists.
>
> (1998:68–69)

It's true that the newspaper reports focused on the women as victims in both senses of the word. On one hand, as brutalized victims of the Ripper but second, viewing them as fallen women, who through various circumstances have taken to prostitution, living in squalor and earning only enough money for a room for the night and alcohol to sustain them.

The Ripper victims themselves have become players in the drama, and the endless discussions from Ripperologists about who the Ripper killed has become a macabre game; women are placed in an acceptable canon of murder, an exclusive club whose membership is based on who killed who.

Megha Anwer's study of the Ripper victim's mortuary photographs suggests that the victims are further objectified by being presented as homogenised tableaus of 'dead flesh' rather than seen as individual women, in particular that of the Ripper's supposed final murder where 'Mary Kelly's brutalized body and disfigured features signal a chilling loss of identity or identification. Her hopelessly anonymous, vanished face becomes infinitely transposable: the defaced prostitute could be literally anyone' (2014:437). Here the women are reduced to props to be displayed in news reports and police crime sheets. L. Perry Curtis also discusses the ways in which the mortuary photographs and news reports reveal very little about the women themselves and say more about the dominance of the male gaze (2001:234). Very little biographical data about the Ripper's victims was provided by the press at the time and instead, even today, much more is written about the list of suspects.

Gareth Steadman-Jones further argues that popular readings of the East End rookeries were 'suffused with all the trappings of a romantic horror story… [a feeling intensified] by a spate of criminal novels' (1984: 179–180). He does concede that romantic as they were, the stories did conceal 'the fear of a hardened senior criminal race of outlaws' (ibid), but the idea that crime was romantic persisted until the brutalities of the Ripper's urban crimes were reported in all their sensationalist detail. To this end, we can observe that the reporting of the Whitechapel murders helped open up a dialogue about conditions in the East End, revealing to a wider audience (both home and overseas) that something was rotten at the heart of Queen Victoria's Empire. A year earlier, 1887 saw Queen Victoria celebrate her Golden Jubilee, and Britain was still considered a great Imperial power, although the 1880s witnessed several riots in London, protesting against working and living conditions. In particular, the violent Bloody Sunday riots of 1887 further demonstrated the ill feeling towards the British monarchy and Government, in Ireland as well as in mainland Britain. The veneer of Victorian life was gradually being peeled away, revealing a different underbelly, both in British society and in fiction.

While Sherlock Holmes would open up and reveal a London that wasn't always so barbaric (there are few details of violent sexual crimes in the Holmes canon), the reporting of the Ripper would also reveal the darker underbelly of the neglected Whitechapel area. Both function, then, to reveal versions of the East End of London. Martin Willis examines the close relationship between Sherlock Holmes and Jack the Ripper further, acknowledging that both 'characters' have subsequently been set up in opposition to one another (2007:144), not only in twentieth-century fiction (such as the films *A Study in Terror* (1965) and *Murder By Decree* (1979), and Michael Dibdin's 1978 novel, *The Last Sherlock Holmes Story*), but also as early as the 1890s, where the author was asked 'to comment on how Sherlock Holmes would have investigated the Whitechapel murders' (ibid). Willis argues that Conan Doyle, indirectly (in terms of the Ripper) provided his readership with the closure they didn't receive from the 1888 murders (2007:156). While Sherlock Holmes was created after the Whitechapel murders, there is little doubt that Conan Doyle was aware of the events and did inform his writing and the audience's appetite for crime and detective fiction. The success of Conan Doyle's detective fiction can, in part, be attributed to the intensified interest in crime and crime reporting during the Whitechapel murders.

Similarly, Bram Stoker's *Dracula*, published in 1897, also draws on the stories published by the press in the late 1880s. If we consider Stoker's preface to the Icelandic edition of *Dracula*, it is clear that the Ripper drama informed some of Stoker's extensive research when preparing for his novel:

> This series of crimes has not yet passed from the memory – a series of crimes which appear to have originated from the same source and which at the time created as much repugnance in people everywhere as the notorious murders of Jack the Ripper.
>
> (Stoker, 1901)

Stoker's novel features women (and men) being terrorized at night, and the allusion to fangs suggests the act of ripping quite clearly. This point is taken up by Nicholas Rance, who argues that the Ripper and Count Dracula can be read as moral avengers who rescue men from bestial women. Rance acknowledges that while virtuous women are saved (Mina is the example), the sexualized women (which can be read as the Whitechapel prostitutes) are destroyed not by Dracula but by, or under the guidance of, Professor Van Helsing. His comparison continues, suggesting that Van Helsing evokes the police and newspaper theory that the Ripper was a medical professional, whose surgical precision was employed when murdering his Whitechapel victims. Rance provides an interesting and persuasive argument in relation to the novel but furthermore, the very suggestion that Dracula was itself informed by the Whitechapel murders further lends itself to the idea that the Ripper has become blurred with other fictional characters, including Spring-Heeled Jack, prior to the 1888 murders and later, the Sherlock Holmes stories. The Ripper in effect entered as fiction and, has remained so to this day.

In this sense, the figure of the Ripper can be seen as an extension of Victorian stage melodrama and the penny dreadfuls (sometimes referred to as penny bloods to alert readers to their violent content) that were popular in the mid-1880s. Drawing on gothic literature, the dreadfuls and the gothic stage melodramas tended towards sensational horror and crime. They were visceral in nature, utilizing garish descriptions or an excess of spectacle to tell their stories. While the stage melodramas relied on visual effects, the penny dreadful magazines utilized dense prose to detail the crimes of their antagonists who, as Springhall remarks, were often seen as anti-heroes as well as outright villains. The image of the Ripper, as a tall, cloaked figure often portrayed as wearing a tall hat, can be traced back to the villainous figures of stage melodrama, often the aristocratic scoundrel who preys on the young, pure woman. So too can the figures of Sweeney Todd, Varney the Vampire and Spring-Heeled Jack be seen as precursors to the image and early understanding of the Ripper, figures who stalked or killed their victims in the East End and were notable for their use of knives or talons (as in the case of Spring-Heeled Jack).

In common with the Ripper, Spring-Heeled Jack was purportedly a real figure, first sighted in 1837. However, his existence as urban myth is cemented in 1840 with the first performance of John Thomas Haines' play *Spring-Heeled Jack, the Terror of London*. Later, in 1863, a forty-part penny dreadful was published, further embedding the figure as a work of fiction. Similarly, Sweeney Todd first appeared in the penny dreadful, *A String of Pearls*, featuring the 'Demon Barber of Fleet Street' who cut the throat of his barbershop clients, who were then used for pie fillings by his partner Mrs Lovett. Again, as with Spring-Heeled Jack and Jack the Ripper, there has been debate as to whether Todd was based on an actual historical figure. Todd's use of the razor certainly offers a precursor to the Ripper's supposed surgical or possibly butcher's knife, further serving to blur the fiction of Sweeney Todd with the figure purportedly responsible for the Whitechapel murders. While Sidney Paget's illustrations for Sherlock Holmes in *The Strand* helped fix the image of the detective wearing a deerstalker hat and provided a visual marker for subsequent portrayals of Holmes on stage and screen, so too did the precursors of the Ripper help fix a visual representation of Jack the Ripper. There are no trusted witness statements or any confirmed actual sightings of the Whitechapel murderer, yet the image of him is based on previous fictional villains and contemporary newspaper illustrations of the nineteenth century. The Ripper is an amalgamation of these figures, becoming a product to be sold, as entertainment as well as a symbol for the anxieties felt within the East End of London.

There is little doubt that the fascination and instant appeal with the nineteenth-century audience – of the Whitechapel murders on the one hand, and the figure of Jack the Ripper on the other – was due to a combination of three significant factors. Most important was the role of the London press, who capitalized on and promoted the sensational aspects of the events of 1888. Building on an industry already preoccupied with crime reporting, particularly violent crime, the murders attributed to a person or persons known as Jack the Ripper simply allowed the industry to exploit the situation further. The longer the crimes remained unsolved, the longer the mystery was allowed to perpetuate; the more it allowed the industry to tap into the anxieties and indeed the ghoulish interests of its readership. Second, in terms of the broader context, London (as the heart of the British Empire) itself was being revealed in a less favourable light. After the Golden Jubilee celebrations of 1887 were dying down, the tensions and fears of an 'outcast London' were beginning to rise to the surface, through riots, protests and by social observers who were keen to make visible that which, for the most part, had remained hidden. While earlier in the century, Charles Dickens had written about a deprived and largely forgotten or ignored London, in a similar way the London press' reporting of the crimes of the Ripper focused a spotlight on the proverbial fog-blurred and slum-ridden streets of East End

London. The press reports and indeed the killings themselves captured a mood and helped articulate the degradation of a localized part of London. The area, as with the figure of the Ripper, became a symbol for the fears and neglect accompanying the sense of industrial decline in Britain. Compare, for example, the exchange in *A Christmas Carol* (1843) between Scrooge and The Ghost of Christmas Present – 'This boy is Ignorance. This girl is Want. Beware them both, and all of their degree; but most of all beware this boy, for on his brow I see that written which is Doom, unless the writing be erased' (Dickens 1994: 57; see Chapter 13 in this volume[1]) – with *Punch* magazine's caption for the illustration 'The Nemesis of Neglect', published on 29 September 1888[2]:

> There floats a phantom on the slum's foul air
> Shaping, to eyes which have the gift of seeing,
> Into the spectre of that loathly lair.
> Face it – for vain is fleeing.
> Red-handed, ruthless, furtive, un-erect,
> 'Tis murderous crime, the nemesis of neglect.

Punch was simply extending Dickens' cautionary tale to accommodate a new symbol of neglect in the form of Jack the Ripper. The Ripper as a symbol was himself drawn from previous fictional incarnations and can be seen to be continuing, and later influencing a tradition of folk-devils who act not only as scapegoats for social problems, but as beacons drawing our attention to wider social troubles. As his secure place in nineteenth-century popular culture makes clear, villains, as much as heroes, act as a barometer for social attitudes, fears and anxieties. The Ripper's identity may forever be concealed, but the myth of the Ripper acts to reveal the state of a place and a time. Jack the Ripper is the composite of fictions and facts, articulating the problems of 1880s London (and more widely, a rapidly changing, industrialized Britain); and acting as a blank tableau to have inscribed upon him the reaction of a people. In this, he will never be caught and never die.

Notes

1 John Leech's illustration 'Ignorance and Want' from *A Christmas Carol* (1843) can be viewed here: www.bl.uk/romantics-and-victorians/articles/the-origins-of-a-christmas-carol

2 John Tenniel's illustration 'The Nemesis of Neglect' from *Punch* magazine (29th September, 1888) can be viewed here: www.bl.uk/collection-items/the-nemesis-of-neglect-from-punch

References

Anwer, Megha (2014) 'Murder in Black and White: Victorian Crime Scenes and the Ripper Photographs' *Victorian Studies*, Vol. 56, No. 3 (Spring 2014), pp. 433–441.

Babylon 5 (1995) 'Comes the Inquisitor' US: Warner Bros Television. 25 October.
Barefoot, Guy (2001) *Gaslight Melodrama: From Victorian London to 1940s Hollywood*. Continuum: London.
Baudrillard, Jean (1994) *Simulcra and Simulation*. University of Michigan Press: Michigan.
Bloom, Clive, Docherty, Brian, Gibb, Jane and Shand, Keith (eds) (1988) *Nineteenth-Century Suspense*. Macmillan: London.
Cohen, Stanley (1973) *Folk Devils and Moral Panics*. Paladin: St. Albans.
Coville, Gary and Lucanio, Patrick (1999) *Jack the Ripper: His Life and Crimes in Popular Entertainment*. McFarland: London.
Crone, Rosalind (2012) *Violent Victorians*. Manchester University Press: Manchester.
Curtis, L. Perry (2001) *Jack the Ripper and the London Press*. Yale University Press: New Haven, CT.
Diamond, Michael (2004) *Victorian Sensation*. Anthem Press: London.
Dickens, Charles (1994) 'A Christmas Carol', *The Christmas Books*. Penguin: London.
Flanders, Judith (2011) *The Invention of Murder*. Harper Press: London.
Rance, Nicholas (2002) '"Jonathan's Great Knife": "Dracula" Meets Jack the Ripper', *Victorian Literature and Culture*, Vol. 30, No. 2 (2002), pp. 439–453.
Springhall, John (1998) *Youth, Popular Culture and Moral Panics*. Macmillan Press: London.
Star Trek (1967) 'Wolf in the Fold' US: CBS Television Distribution. 22 December.
Steadman-Jones, Gareth (1984) *Outcast London*. Penguin: London.
Stoker, Bram (1901) Reprinted in Leatherdale, Clive (2006) *Dracula Unearthed*. Desert Island Books: Brighton.
Storey, John (ed) (2009) *Cultural Theory and Popular Culture: A Reader*, 4th edition. London: Routledge.
Time After Time (1979) Directed by Nicholas Meyer [Film]. USA: Warner Bros.
Warwick, Alexandra and Willis, Martin (eds) (2007) *Jack the Ripper: Media, Culture, History*. Manchester University Press: Manchester.

Index

Note: Page numbers in *italic* indicate figures and tables.

For Product Safety Concerns and Information please contact our EU representative GPSR@taylorandfrancis.com
Taylor & Francis Verlag GmbH, Kaufingerstraße 24, 80331 München, Germany

www.ingramcontent.com/pod-product-compliance
Lightning Source LLC
LaVergne TN
LVHW021124300926
849160LV00010B/857
* 9 7 8 1 1 3 8 8 5 4 9 1 8 *